ANNA L. GIRARD

Essays on Scent Marketing:

Effects of Scented Indoor Environments on Customers and Employees

FGM-Verlag
Verlag der FGM Fördergesellschaft Marketing e.V.
an der Ludwig-Maximilians-Universität München

Schriftenreihe SCHWERPUNKT MARKETING
Band 87

Herausgeber: Univ.-Prof. Dr. Paul W. Meyer †/Univ.-Prof. Dr. Anton Meyer

Girard, Anna L.:

Essays on Scent Marketing:

Effects of Scented Indoor Environments on Customers and Employees

FGM-Verl., Verl. der Fördergesellschaft Marketing e.V., 2015

(Schriftenreihe Schwerpunkt Marketing; Bd. 87)

Zugl.: München, Univ., Diss., 2015

ISBN 978-3-945496-07-7

Die Dissertation wurde unter dem Titel „Essays on Scent Marketing: Effects of Scented Indoor Environments on Customers and Employees" eingereicht.

Datum der Promotionsabschlussberatung: 13. Mai 2015

Referent: Univ.-Prof. Dr. Anton Meyer
Koreferent: Univ.-Prof. Dr. Ingo Weller

ISBN 978-3-945496-07-7

Essays on Scent Marketing:
Effects of Scented Indoor Environments on Customers and Employees

Inaugural-Dissertation zur Erlangung des Grades
Doktor oeconomiae publicae (Dr. oec. publ.)
an der Ludwig-Maximilians-Universität München

vorgelegt von
Anna Girard, B.Sc., MBR
im Jahr 2014

Referent: Univ.-Prof. Dr. Anton Meyer

Koreferent: Univ.-Prof. Dr. Ingo Weller

Promotionsabschlussberatung: 13. Mai 2015

EDITOR'S PREFACE

It doesn't matter where you go: one cannot hide from scents that swirl around us, as it is nearly impossible not to breathe and smell your environment. This anatomical uniqueness makes the sense of smell to one of the most powerful of our senses; especially since it is also known as an emotional sense with a direct link to our brain. Therefore, scent marketing – or the purposeful diffusion of pleasurable scents in commercial environments – is considered a highly powerful Marketing tool, which is already used by several companies, like Singapore Airlines, Samsung, or Abercrombie & Fitch. However, it is important that companies applying ambient scents develop a deeper understanding of the underlying processes of olfactory perception, in order to fully leverage its potential.

So far, there is a lack of research especially on the impacts of ambient scents in service environments, and more importantly on the temporal structure of their influence. Since companies typically use scents over a longer period, once they get implemented, especially the long-term effects are crucial from a managerial perspective. Surprisingly, there exist no insights in neither scent nor service research so far. In this regard, Anna Girard's dissertation contributes to service and scent research in multiple ways and not only provides highly relevant insights and direction for further research, but also provides an impactful managerial guide on how to use ambient scents in marketing practice. This thesis provides conceptual and empirically well-grounded insights presenting three distinct articles: The first article conceptualizes a theoretical framework that holistically describes the process of olfactory stimulation and its impacts on customers and employees in service environments. Based on these considerations the second article empirically investigates the long-term influence of ambient scents on service customers. Finally, to provide a holistic picture, the third article investigates ambient scent's effects on employees over time.

This dissertation continues the research tradition of my institute in service research and also opens new paths for research through its interdisciplinary approach. It is not only an interesting read, but can also serve as guide for those interested in creating a superior service experience for their brand, by leveraging the so far largely neglected sense of smell.

Univ.-Prof. Dr. Anton Meyer   Munich, May 2015

AUTHOR' PREFACE

No text – No explanation – Smell and think.

Oswaldo Marciá

Writing a dissertation is just about facing yourself – mostly all alone and sometimes with a little help of your colleagues and friends. Thus, there are some people I want to thank:

- Prof. Meyer for giving me the opportunity to realize my dissertation project

- Prof. Weller for forcing me to find an explanation for everything

- Marko for being my unexpected mentor through the challenges in academia

- Bernd for supporting the idea of scent marketing in business

- RMG for his endless patience and never ending supply of scent cartridges

- My seminar students for getting up so early and consuming a lot of coffee

- Jan & Marcus for turning colleagues into friends

- Johan for activating my passive voice

- My Mom for inspiring me by writing "Pertussis in Adults"

- And finally Marc for being my love and my toughest reviewer

Anna L. Girard   Nuremberg, May 2015

CONTENT

FIGURES

TABLES

APPENDICES

Chapter C:

Chapter D:

ABBREVIATIONS

adj.	adjusted
AL	adaptation level
ANOVA	analysis of variance
B	behavior
Bonf	Bonferroni
ChemG	Chemikaliengesetz
CNS	central nervous system
COO	Chief Operating Officer
df	degrees of freedom
df_M	degrees of freedom for the effect of the model
df_R	degrees of freedom for the residuals of the model
E	environment
e.g.	exempli gratia (for example)
f	function
F	F-test
FET	Fisher's exact test
fMRI	functional magnetic resonance imaging
Fri	Friday
GefahrstoffV	Gefahrstoffverordnung
H	hypothesis
IAQ	indoor air quality
i.e.	id est (that is)
k	number of levels of the within-subjects factor
KMO	Kaiser-Meyer-Olkin
LFGB	Lebensmittel- und Futtermittelgesetzbuch
M	mean
max	maximum
MBR	Master of Business Research
min	minimum
mio	million
Mon	Monday
MSA	measure of sampling adequacy
N	number (sample size)
N/A	not available
No.	number

n.s.	not significant
P	person
p	p-value
p.	page
PAQ	perceived air quality
PCA	principal component analysis
pp.	pages
r	correlation coefficient (effect size)
R^2	coefficient of determination
REACH-Regulation	European Union regulation concerning the registration, evaluation, authorisation & restriction of chemicals
RQ	research question
rANOVA	repeated-measures ANOVA
SBS	sick building syndrome
sig.	significant
t	t-test
Thu	Thursday
t_n	time point n
Tue	Tuesday
U	Mann-Whitney U-test
US	United States (of America)
V	Cramer's V (effect size)
W	wave
Wed	Wednesday
α	Cronbach's α
Δ	delta (difference)
ε	Greenhouse-Geisser correction
η^2	eta square (effect size)
η_p^2	partial eta square (effect size)
σ	standard deviation
χ^2	chi-square-test
χ^2_F	Friedman's related-samples ANOVA
ω^2	omega square (effect size)
€	Euro
$	Dollar

A. INTRODUCTION

"We see only when there is light enough,
taste only when we put things into our mouths,
touch only when we make contact with someone or something,
hear only sounds that are loud enough.
But we smell always and with every breath.

Cover your eyes and you will stop seeing,
cover your ears and you will stop hearing,
but if you cover your nose and try to stop smelling,
you will die." (Ackerman, 1990, p. 6)

1   The Relevance of Scent Marketing

Almost everything emits its individual scent to the environment, including people, plants, animals, and materials (Hatt, 2007): "Smells coat us, swirl around
us, enter our bodies, emanate from us. We live in a constant wash of them"
(Ackerman, 1990, p. 7). Thus, scents are everywhere, and since humans need
to breathe more than 23,000 times per day, we can hardly avoid them
(Ackerman, 1990; Bradford & Desrochers, 2009). Already Laird (1935) concluded that "[t]he sense of smell determines much more of our behavior than we like
to admit, or than we consciously realize" (p. 126). Since then, research has
demonstrated that scents and the human sense of smell do have a tremendous
impact on peoples' emotions, cognitions, and even behavior (M. Girard, 2015;
Olahut, 2013; Teller & Dennis, 2012).

This is why, in the late 1990s, companies started to leverage *scent marketing,*
which utilizes our sense of smell in order "to set a mood, promote products [or
services,] or position a brand" (Vlahos, 2007, p. 70). "Scent marketing relies on
the neuropsychological processing of olfactory stimuli in the human brain"
(Emsenhuber, 2011, p. 350), and relates to two main areas of commercial scent
application: Ambient scents and brand scents (Elejalde-Ruiz, 2014). An *ambient
scent* is defined as "a general odor which does not emanate from a product but
is present as part of the (…) environment" (Bradford & Desrochers, 2009, p.
143), and infuses an indoor space with a pleasurable smell (Elejalde-Ruiz,
2014). A *brand scent* (also referred to as signature scent) (Elejalde-Ruiz, 2014)
is defined as a pleasant scent that is specifically designed for a brand and is

intended to communicate the brand identity via the olfactory sense (Girard et al., 2013), "like an olfactory logo" (Elejalde-Ruiz, 2014).[1]

Back in 2005, 99% of all company communication was concentrated on only two of our five senses: Vision and hearing (Lindstrom, 2005b). Since then, the sense of smell has received steadily increasing awareness among marketers (Klara, 2012), and the scent marketing industry is growing swiftly (Elejalde-Ruiz, 2014). In 2008, roughly 35% of Fortune 500 companies were estimated to use scent marketing already (Bell, 2007). Still, "[s]cent is grossly underutilized by brand builders," says Donna Sturgess, President of the New York marketing consultancy Buyology (Klara, 2012).

All publicly available estimates of spendings in the scent marketing industry are based on statements by the Scent Marketing Institute, "the leading authority on scent-supported marketing" (Scent Marketing Institute, 2011), which has declared itself the "official voice of the scent marketing industry" (ScentWorld Marketing, 2013) (see Figure 1).

Figure 1: Growth Projections of the Global Scent Marketing Industry[2]

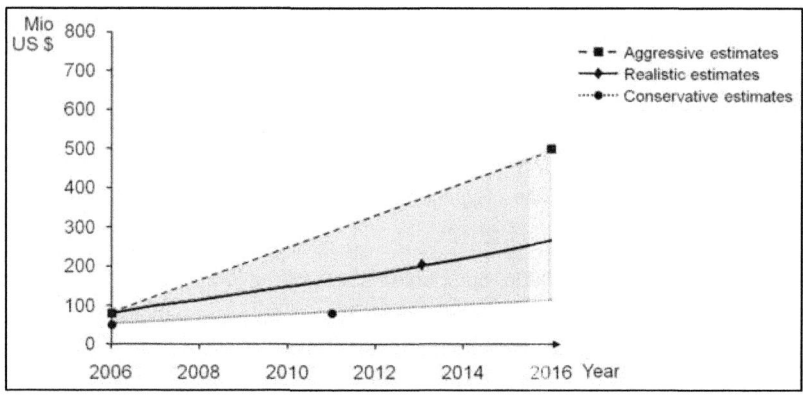

Harald Vogt, Founder of the Scent Marketing Institute, estimated that US$50 to $80 million were spent on scent marketing in 2006 (Vlahos, 2007). More recent data has been provided by Jennifer Dublino, the Scent Marketing Institute's

---

[1] While companies might use an ambient scent, which is not necessarily also a brand scent, and vice versa, there are examples like Abercrombie & Fitch that also use their proprietary brand scent as an ambient scent.
[2] Own illustration based on data from Elejalde-Ruiz (2014); Ravn (2007); Sutton (2011); Vlahos (2007).

COO: She estimated worldwide industry revenues at $80 to $100 million a year in 2011 (Sutton, 2011). For 2013, her industry revenue projections were $200 million, which is forecasted to further increase by an predicted annual growth rate of roughly 10% per year (Elejalde-Ruiz, 2014). More aggressively, Harald Vogt has predicted that scent marketing would reach at least $500 million by 2016 (Ravn, 2007; Vlahos, 2007).

Overall, scent marketing "is a huge trend", as Eric Spangenberg, Marketing Professor from Washington State University, acknowledges (Klara, 2012). This is why scent marketing was not only proclaimed an important business trend to watch by the management journal *Advertising Age* for 2007 (Thomaselli, 2006), but also a phenomenon worth investigating by academia (Klara, 2012). To date, the sense of smell has been under-researched, especially compared to the other senses, even though "this has been changing in recent years, and we're starting to see it be introduced in academia. The science is clear: Scent has incredible potential," concludes Donna Sturgess (Klara, 2012). This might be why Grewal and Levy (2009) considered the influence of scent as an environmental cue to be one of the key emerging issues in *Journal of Retailing* between 2002 and 2007.

However, many aspects of scent research are still "in its infancy and many elements of the smell phenomenon remain unexplained" (Emsenhuber, 2011, p. 350). The German Federal Environment Agency therefore recommends intensifying university research on the effects of (ambient) scents (Umweltbundesamt, 2006a). Marta Tafalla, Professor in Philosophy at the University of Barcelona, who suffers from anosmia (an inability to smell at all), has called for olfaction to "be studied in a biological and medical way and also psychologically and anthropologically, and from every possible discipline" (Tafalla, 2013, p. 1296), in order to advance knowledge. "But the scientific community has already achieved something very important putting olfaction on the map of science, after it was so neglected for centuries. And it is really important to continue" (Tafalla, 2013, p. 1296).

This dissertation seeks to make a scientific contribution in this regard.

2   Objectives and Content of this Thesis

The general objective of this dissertation thesis is to contribute to current knowledge on scent marketing research, specifically the impact of scented environments (ambient scents) on customers and employees.

Following the initial introduction (Chapter A), this cumulative dissertation consists of three independent and self-contained articles written to be submitted separately to international scientific journals.[3] Owing to a scientific journal writing style, I refrained from defining common marketing terms and explaining statistical procedures in detail. Since each paper targets one specific journal, their styles diverge somewhat.

As all three articles belong to the same research area – of scent marketing – the reader will find some redundancies, especially in the sections of scent characteristics and scent perception. However, all three papers have different perspectives, goals, and theoretical foundations, which will be now outlined.

Paper 1: The Scentscape – An Integrative Framework Describing Ambient Scents in the Servicescape (Chapter B)

The first – conceptual – article develops an integrative framework of how ambient scents affect individuals present in a service environment (customers and employees). In the past decades, several empirical papers have investigated ambient scents' effects on customers and, to some extent, on employees. However, we lack a theoretically motivated framework that provides an overview and addresses ambient scents' role in the physical surroundings of service settings. Hence, paper 1 seeks to answer the following research questions (RQ):

RQ1. What impacts do pleasant ambient scents have present in a service environment?

RQ2. Which factors influence ambient scents' impacts on individuals in a service environment?

---

[3] Except for paper one, which is a co-authored paper, I use "we" instead of "I" also for single-author papers because this is common use in scientific journals. "We" is thus not an indication of co-authors unless specifically stated.

This conceptual paper develops an integrative framework – the *scentscape*[4] – that describes ambient scents in service environments: The scentscape holistically describes the process of olfactory stimulation and its impacts on customers and employees in service environments by extending Bitner's servicescape model (1992) and combining it with Gulas and Bloch's (1995) model of the influence of ambient scent. The scentscape illustrates different sources of scents and their stimulation of customers and/or employees present in a servicescape. The model also outlines the processes of scent perception as well as individuals' internal reactions and behavioral responses evoked by scent exposure. For academics, the scentscape model gives a visual overview of a range of influencing factors that should be considered when investigating ambient scent's effects; it also triggers future empirical research. For practitioners, the scentscape model provides a framework to guide managers concerning relevant issues and questions that should be raised and answered before introducing ambient scents into a service environment.

Paper 1, which was co-authored with Marc Girard and Anna-Caroline Suppin, represents the basic framework of this dissertation and will be referenced in the following two empirical articles as (Girard et al., 2015).

Paper 2: Are You on the Right Scent? Ambient Scents' Short- and Long-term Effects on Customers in a Servicescape (Chapter C)

Paper 2 empirically investigates the temporal structure of ambient scent effects on service customers. To date, scent research has focused only on one-time scent exposures, while in reality, customers might frequently visit a specific service provider, and the effects of repeated exposures might differ from short-term one-time exposure. Thus, paper 2 seeks to answer the following questions:

RQ1. What are the short-term effects of one-time exposure to ambient scents on customers in a servicescape?

RQ2. What are the long-term effects of repeated exposure to ambient scents on regular customers in a servicescape?

---

[4] The term „Scentscape" is a registered trade mark (word mark) held by Christoph Oldendorf with the register number 30702936 of the Deutsches Patent- und Markenamt.

RQ3. What are aftereffects on regular customers after the removal of a longitudinally diffused ambient scent in a servicescape?

Based on optimal arousal theory, paper 2 derives three hypotheses, which were empirically tested with a controlled field experiment using a within-subjects design over four months (9 waves) with a customer panel (N = 35) in a German public transportation service company. It applies a pretest/posttest control group design to reliably measure the development of the dependent variables over time. The article also includes interviews with 10 panel participants to gather additional qualitative information and insights for the interpretation of our empirical findings. Overall, to our best knowledge, the paper represents the first longitudinal investigation of the scent effects temporal structure in the marketing discipline.

Paper 3: The Impacts of Ambient Scents in the Workplace: A Qualitative Investigation (Chapter D)

Paper 3 empirically investigates the impacts of ambient scents on employees in a workplace. Research on the indoor environmental quality of workplaces has demonstrated that poor air quality has a considerable negative effect on employees. One possibility to enhance indoor air quality is to diffuse pleasurable ambient scents. However, little research has investigated ambient scent's effect on employees in their workspace. To date, there are, to our best knowledge, few empirical findings in real work environment settings. Hence, based on the theoretical considerations of field theory, paper 3 seeks to deliver first qualitative insights into ambient scent's effects in the workplace by answering following questions:

RQ1. Does the introduction of a pleasurable ambient scent in the workplace lead to emotional, cognitive, and behavioral responses of employees?

RQ2. Is there a difference between self-reported and observed scent effects on employees?

RQ3. Does an employee's perception of an ambient scent vary over time?

The research questions are answered by using a qualitative diary with seven employees over five workdays in a scented work environment. Overall, the article concludes in that olfactory cues present in workspaces do affect internal responses and behavior of employees. Thus, paper 3 provides first qualitative insights into ambient scent's effects in the workplace on employees, its temporal structure, and which considerations should be made by managers before its introduction.

Since each paper provides a separate conclusion, including a discussion of the results, relevant management implications, limitations and implications for further research, the thesis ends with a general discussion of the main findings and measures taken to safeguard good scientific practice, as well as a final conclusion (Chapter E).

Figure 2 provides a visual overview of the thesis.

Figure 2: Overview of Dissertation Thesis[5]

Essays on Scent Marketing: **Effects of Scented Indoor Environments on Customers and Employees**		
A. Introduction: Relevance of Scent Marketing & Objectives of the Thesis		
B. The Scentscape – An Integrative Framework Describing Ambient Scents in the Servicescape Conceptual Paper	**C.** Are You on the Right Scent? Ambient Scents' Short- and Long-term Effects on Customers in a Servicescape Empirical Paper (Field Experiment & Qualitative Interviews)	**D.** The Impact of Ambient Scents in the Workplace: A Qualitative Investigation Empirical Paper (Qualitative Diaries)
E. General Discussion & Conclusion		

---

[5] Own illustration.

B. THE SCENTSCAPE – AN INTEGRATIVE FRAMEWORK DESCRIBING AMBIENT SCENTS IN THE SERVICESCAPE

MARC GIRARD, ANNA L. GIRARD, ANNA-CAROLINE SUPPIN[6]

---

[6] A later version of this paper is published in Girard, M. et al., (2016) "The Scentscape: An Integrative Framework Describing Scents in Servicescapes," Journal of Business Market Management, 9 (1), 26.

0   Abstract

The systematic use of scents as a marketing instrument is an intensifying trend in service companies that is accompanied by increasing research attention. By extending Bitner's servicescape model (1992) and combining it with Gulas and Bloch's (1995) model of the influence of ambient scent, this article develops the scentscape – an integrative framework that holistically describes the process of olfactory stimulation and its impacts on customers and employees in service environments. It illustrates different sources of scents and their stimulation of customers and/or employees present in a servicescape. The model also provides an overview of the processes of scent perception as well as individuals' internal reactions and behavioral responses. We propose a future research agenda organized around the elements of our conceptualization. For marketers considering the use of ambient scents in specific service settings, we derive a set of assisting implications. Even if no ambient scents are deliberately introduced into a servicescape, the scentscape illustrates that the olfactory situation in the environment will still influence individuals through their perceived air quality. We therefore argue that active indoor air quality management should be a key task for service managers, in order to avoid the potential negative impact of an unpleasant olfactory experience.

Keywords: scent, servicescape, consumer behavior, employee, air quality

Acknowledgments: The research on which this article is based is part of the first and second authors' dissertation projects, as well as the third author's Bachelor thesis (under the first author's supervision) at the Ludwig-Maximilians-University in Munich. The authors acknowledge valuable comments by Prof. Dr. Anton Meyer, Prof. Dr. Marko Sarstedt, Prof. Mary Jo Bitner, Ph.D., Dr. Silke Bartsch, and three anonymous reviewers of the *Journal of Service Research*.

1 Introduction

The systematic use of scents is an intensifying trend in service companies (Bell, 2007; Klara, 2012). Meanwhile, marketers intentionally incorporate scents into locations' interior design in a variety of settings, ranging from retail (e.g., Abercrombie & Fitch created a brand scent for use in stores, on textiles, and employee perfume), to transportation services (e.g., Singapore Airlines holistically integrates a signature scent into its communication concept as perfume, ambient scent, and refreshing tissues), to hotel businesses (e.g., the Starwood Group created a unique fragrance for each hotel chain that mainly used in the lobbies), and many other areas (Bell, 2007; Goldkuhl & Styvén, 2007).

Using scents in a service provider's physical environment might generally be related to various managerial objectives ranging from covering bad smells present in the service delivery location, to enhancing a service brand's recall value, to increasing customers' service experiences and employee performance – mostly with the ultimate goal of increasing sales (Goldkuhl & Styvén, 2007; Lindstrom, 2005a). An understanding of scents' specific characteristics and effects is therefore crucial to help companies to effectively and cautiously leverage the power of the sense of smell in service environments.

The marketing literature has long acknowledged that ambient scents, as part of the physical environment, affect individuals (Lam, 2001), and has considered its influence predominantly in retail or laboratory settings (e.g., Hirsch, 1995; Laird, 1935; Mattila & Wirtz, 2001; Morrin & Chebat, 2005; Morrison et al., 2011; Spangenberg et al., 1996). Over the past decades, empirical research has analyzed a variety of isolated variables that can potentially be influenced by scents (see Table 1).

Table 1: Overview of Investigated Scent Responses[7]

Scent effect	Dependent variable	Study
Physiological activation	Skin conductance responses	Møller & Dijksterhuis (2003)
	Brain activity	Masago et al. (2000); Owen & Patterson (2002); van Toller et al. (1993)
	Blood pressure	Heuberger et al. (2001, 2004); Höferl et al. (2006)
	Pupillary responses	Schneider et al. (2009)
	Sleep quality	Badia et al. (1990); Goel et al. (2005); Raudenbush et al. (2003)
	Pain	Marchand & Arsenault (2002); Villemure et al. (2003)
Emotional responses	Emotions	Asmus & Bell (1999); Knasko (1992); Mattila & Wirtz (2001); Michon et al. (2005); Mitchell et al. (1995); Morrin & Chebat (2005); Morrison et al. (2011); Rotton (1983); Schifferstein et al. (2011), Spangenberg et al. (2005); Teller & Dennis (2012)
	Mood	Baron (1990); Baron & Thornley (1994); Gilbert et al. (1997); Knasko (1993b, 1995); Ludvigson & Rottman (1989); Mitchell et al. (1995); Warm et al. (1991)
	Emotional memories	Baeyens et al. (1996); Epple & Herz (1999); Herz & Cupchik (1992); Laird (1935); Lehrner et al. (2000); Lehrner et al. (2005); Robin et al. (1999)
Cognitive responses	Evaluation of environment	Baron (1990); Baron & Bronfen (1994); Michon et al. (2005); Morrin & Chebat (2005); Parsons (2009); Schifferstein et al. (2011); Spangenberg et al. (1996, 2005, 2006)
	Evaluation of persons present in environment	Baron (1981, 1983, 1986); Fiore & Kim (1997); Li et al. (2007); Maille (2006); Mattila & Wirtz (2001); McGlone et al. (2013)
	Evaluation of service quality	Girard et al. (2013); Maille (2006); McDonnell (2007); Morrin & Chebat (2005)
	Evaluation of service experience	Girard et al. (2013)
	Evaluation of brand	Girard et al. (2013); Mani (1999); Morrin & Ratneshwar (2000, 2003)
	Time perception	Maille (2006); Spangenberg et al. (1996, 2006)
	Overall satisfaction	Mattila & Wirtz (2001); Morrison et al. (2011)
	Self-evaluation	Baron (1990); Gilbert et al. (1997); Knasko (1992, 1993)
	Stress perception	Baron & Bronfen (1994); Baron & Thornley (1994)
Behavioral reponses	Behavioral intention	Fiore et al. (2000); Herrmann et al. (2013); Spangenberg et al. (1996, 2005, 2006)
	Approach / avoidance behavior	Asmus & Bell (1999); Doucé et al. (2013); Guéguen & Petr (2006); Maille (2006); Mattila & Wirtz (2001); Schifferstein et al. (2011)
	Buying behavior	Chebat et al. (2009); Guéguen & Petr (2006); Hirsch (1995); Jacob et al. (2014); Morrin & Chebat (2005); Spangenberg et al. (2006)
	Physical performance	Barker et al. (2003); Ho & Spence (2005); Raudenbush et al. (2002); Sakamoto et al. (2005); Warm et al. (1991)
	Mathematical tasks	Degel & Köster (1999); Gilbert et al. (1997); Knasko (1993); Ludvigson & Rottman (1989)
	Linguistic tasks	Baron & Bronfen (1994); Baron & Thornley (1994); Degel & Köster (1999); Habel et al. (2007); Herrmann et al. (2013); Knasko (1993); Nordin et al. (2013); Rotton (1983)
	Creativity tasks	Degel & Köster (1999); Knasko (1992)
Social interaction	Quantity	Doucé et al. (2013); Zemke & Shoemaker (2007, 2008)
	Quality	Baron (1990, 1997); Baron & Thornley (1994)

---

[7] Own illustration.

In fact, research on the influence of ambient scents has predominantly focused on the customer perspective, and few studies have investigated scents' impacts on work-related contexts. However, a typical service environment affects not only a service provider's customers, but also the employees involved in the service delivery process (Bitner, 1990, 1992; Parish et al., 2008). The introduction of ambient scents will thus affect the perception and satisfaction of both customers as well as employees and will become an integral part of their experiences and interactions (Berry & Lampo, 2000; Bitner, 1992; Clarke & Schmidt, 1995). Interestingly, there are also very few explicit studies on scents in service settings that account for the specifics of service environments and service delivery (Maille, 2006; McDonnell, 2007), despite the increased importance of external environmental cues due to a service's intangibility (Zeithaml, 1981).

Overall, there is increasing acceptance among both practitioners and academics that pleasant ambient scent can positively affect individuals.

In conclusion, although there has recently been research, many academics have called for additional investigation of the potential of scents as a controllable marketing instrument (Bosmans, 2006; Morrin & Ratneshwar, 2000; Zomerdijk & Voss, 2010): "One of the least-understood variables in an environment's ambient conditions is ambient scent" (Zemke & Shoemaker, 2007, p. 929). Owing to the very fragmented research (see Table 1, and Teller & Dennis, 2012), we especially lack a theoretically motivated framework that addresses ambient scents' role in the physical surroundings of service settings. Thus, both a theoretical as well as a managerial need exists for a more in-depth understanding of scents in service environments. This paper seeks to contribute to the emerging body of research on scents in service settings by addressing the following central research questions:

RQ1. What impacts do pleasant ambient scents have on individuals present in a service environment?

RQ2. Which factors influence ambient scents' impacts on individuals in a service environment?

This article merges existing theories and empirical findings from various disciplines into a framework that describes how ambient scents affect both customers and employees in a service environment. We build on the structure of Bitner's servicescape model (1992) and combine it with the specific findings of Gulas and Bloch's (1995) model of the influence of ambient scents. Further, we will review scent-related research and will integrate findings from service literature to develop the scentscape as an *integrative framework* for scents in service environments (MacInnis, 2011). The scentscape holistically describes the process of olfactory stimulation and its impacts on customers and employees, while considering specific characteristics and relevant moderators in service settings. This is a step towards enabling organizations to evaluate the influencing factors and to build an understanding of the potential effects of ambient scents, leading to a more effective integration of olfactory stimuli into an organization's marketing strategy. Finally, we derive 11 major insights and discuss key managerial as well as research implications.

2   Underlying Conceptual Models

As a starting point, we will first shortly describe the two basic models, on which or scentscape framework is based: The model of the influence of ambient scent by Gulas and Bloch (1995) and the servicescape model by Bitner (1992, 2000).

2.1   The Model of the Influence of Ambient Scent

Ambient scents are defined as "scents that are present in the environment and do not emanate from a specific product" (Bosmans, 2006, p. 32). Scientific publications in marketing on ambient scents' influences often refer to the conceptual model of Gulas and Bloch (1995), which is based on Mehrabian and Russell's (1974) general model of environmental psychology. The *model of the influence of ambient scent on consumer responses* is used especially in the context of retailing (e.g., Mattila & Wirtz, 2001; Morrison et al., 2011; Spangenberg et al., 1996).

The model postulates that a scent needs to be perceived consciously in order to affect consumer behavior. Depending on the individual olfactory scent acuity (sensitivity to scents) and moderated by personal scent preferences, scent perception then leads to internal affective consumer responses (e.g., change of mood). In turn, these reactions induce approach or avoidance behavior towards an individual's environment. Gulas and Bloch (1995) argue that individual characteristics such as gender and age affect an individual's scent acuity, which in turn determines whether or not a specific scent is perceived consciously. However, besides individual characteristics, physiological predispositions and past experiences with scents influence preferences for specific scents, which then moderate a customer's affective response. Finally, other atmospheric elements (e.g., music) as well as a scent's congruence with other features of the environment (e.g., colors) also play a moderating role. Gulas and Bloch (1995) emphasize that a surrounding has multidimensional facets; therefore, elements of an environment should not be regarded separately and in isolation.

Gulas and Bloch's model undoubtedly helps explain the 'black box' of scent perceptions and their effects on consumers (Davies et al., 2003). However, the model simplifies some aspects of scent perception, processing and reactions, which makes further adaptations necessary. To further strengthen the model's explanatory power, we will discuss potential weaknesses in detail (e.g., scent as one-dimensional input factor, unconscious perception of scents, additional moderators in scent perception, and physiological and cognitive reactions to scents) and integrate additional findings from multidisciplinary scent research into our final scentscape framework. Prior to that, we introduce the servicescape model as basis, specifically explaining environmental influence and effects in service environments.

2.2 The Servicescape Model

Bitner's (1992) model of the *servicescape* – defined as "the immediate physical and social environments surrounding a service experience, transaction or event" (Bitner, 2000, p. 48) – is one of the most salient models in service research. According to the servicescape model, ambient scents are specific environmental cues that are an integral part of a service environment (among many others), and thus affect its holistic perception. Influenced by individual moderators or situational factors, olfactory stimuli might cause emotional and cognitive internal reactions within customers and employees and therefore affect their approach/avoidance behavior as well as possible social interactions (Bitner, 1992, 2000). In turn, their behavior or visible emotions (evoked by scents) can affect social interactions as well as perceptions of the servicescape (Rosenbaum & Massiah, 2011). Bitner (1992) notes: "The ability of the physical environment to influence behaviors and to create an image is particularly apparent for service businesses" (p. 57).

Bitner's servicescape model provides many additional aspects that were not considered in Gulas and Bloch's model of ambient scent so far – as air quality, physiological and cognitive scent responses of customers and employees, and ambient scent's influence on social interaction – this is why we now merge the two perspectives.

3   Integrative Framework for Scents in Servicescapes: The Scentscape

Bringing the model of the influence of ambient scent and the servicescape together, we developed the *scentscape* as an integrated framework that describes scents in a service environment (see Figure 3). The scentscape illustrates different sources of scents and their stimulation of customers and/or employees present in a servicescape. The scentscape also provides an overview of the processes of scent perception as well as individuals' evoked internal responses and behavioral reactions. We derive 11 major insights for service researchers and practitioners.

Figure 3: The Scentscape Model[8]

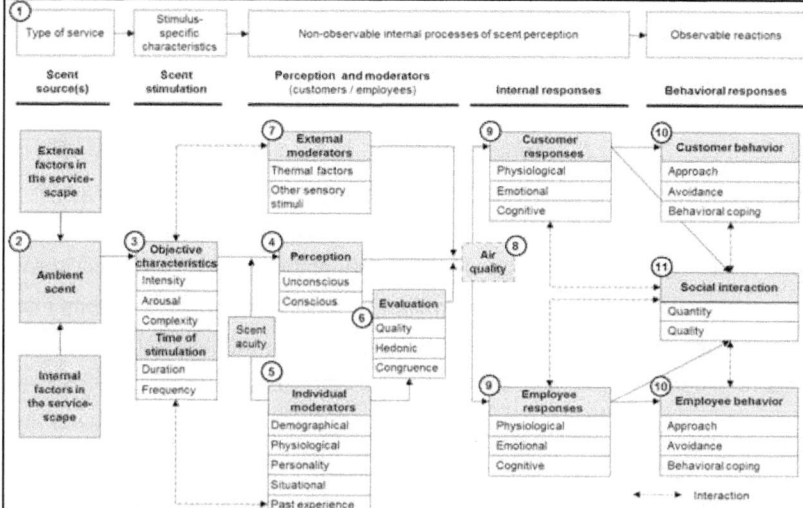

[8] Own illustration.

Insight 1: The importance of ambient scents depends on the **service type**: the lower the extent of search qualities of a service, the higher the importance of scents.

Intangibility usually complicates the evaluation of a service. In contrast to products, most services lack specific search qualities, that is, they typically possess only a few characteristics a customer can evaluate prior to their utilization (e.g., design or packaging). Instead, experience qualities – those features that can only be assessed during or after use (e.g., entertainment in a theater), along with credence qualities, which cannot be evaluated even after the service utilization (e.g., legal advice), are distinct aspects of most services (Zeithaml, 1981). Hence, the evaluation simplicity largely depends on the service type: "While consumers may find it easy to evaluate the performance of everyday services (...) prior to consumption, they may find it impossible to judge those performed by professionals and specialists" (Zeithaml, 1981, p. 187), owing to a high extent of credence qualities. Due to this intangibility, the importance of external environmental cues – such as ambient scents – in customers' evaluation and determining of a service's quality generally increases (Baker et al., 2002; Meyer, 1991; Zeithaml, 1981), especially in service businesses, where the actual service experience and trust in the service delivery appropriateness are distinct (e.g., an amusement park, or a medical examination). Thus, we assume that, the lower the extent of search qualities of a service, the higher the importance of scents as external environmental cues, as "a proprietary scent can 'tangibilize' a company's service" (Zemke & Shoemaker, 2007, p. 937).

Insight 2: Besides actively introduced pleasurable ambient scents, other **internal and external scent sources** in the servicescape influence individuals during service delivery.

Contrary to product-specific scents, ambient scents do not emanate from a specific product, but can be viewed as a manageable environmental cue, for instance, via a service location's air conditioning system (Bosmans, 2006). However, not only directly manageable olfactory cues should be taken into account in service settings; Bitner (1992) subsumes these additional olfactory cues as air quality in her servicescape model. To specify aspects of air quality, we distinguish between internal and external scent sources. During a service delivery process, either the customer or the customer's object is directly involved (Meyer, 1996). Therefore, olfactory emissions of external individuals or external objects (e.g., the customer's dog at the veterinarian) should be considered, as they might interact with an ambient scent actively introduced into the servicescape by managers. Furthermore, other internal factors (besides the ambient scent) can influence the olfactory situation in a servicescape, such as employee emissions or smelling objects used during the delivery process (e.g., shampoo at a hairdresser). For instance, while a grocery store with a butcher needs to account for the smell of meat as an integral internal factor of the scentscape, the employees themselves as smelling subjects must also be taken into account. We integrate these additional internal and external scent sources as relevant further olfactory stimuli into our model.

The most important aspect for managers is probably whether or not the company is able to control these internal and external emission sources. As most of them are hardly controllable (e.g. customers' perfume or sweat), companies must be aware of and optimally take neutralizing countermeasures to avoid undesired interferences with a pleasurable ambient scent.

Insight 3: Scent is a complex and multi-dimensional cue – its **objective characteristics** comprise its intensity, arousal potential, and complexity, as well as the duration and/or frequency of scent exposure.

A scent's *intensity*, which relates to the perceived strength of a smell impression, depends on the specific concentration of an olfactory stimulus and determines whether or not a specific scent is perceived consciously or unconsciously.[9] Scent perception sensitivity thresholds vary greatly between individuals (Binder et al., 2009). To generalize individual thresholds, collective thresholds are said to be reached if more than 50% of subjects perceive or recognize a specific scent (Mücke & Lemmen, 2010).

A scent's direct influence on the central nervous system (CNS) as relaxing or energizing can be characterized based on its *arousal potential* – independent of the level of consciousness of its perception (Jellinek, 1996; Pfaff, 2006). Different scents have distinct activation potentials, which are associated with corre-

---

[9] Also referred to as subconsciously or subliminal.

sponding levels of vigilance (Gould & Martin, 2001; Heuberger et al., 2001): For instance, the odor of roses has a relaxing effect, whereas that of lemons is said to be arousing. Therefore, depending on the intention of scenting a specific servicescape, the stimulus should be properly selected, and its arousal level should be tested prior to implementation (e.g., a relaxing rose scent seems appropriate for a spa, but inappropriate for a fitness center).

A scent's *complexity* depends on how many components a scent consists of. Herrmann et al. (2013) recently transferred scent complexity from food complexity as distinct scent characteristic, and argue that simple stimuli can be processed by individuals more easily (Schwarz, 2004). The authors show, in several empirical studies, that mono-scents consisting of only one component (e.g., lemon) induce significantly better results in linguistic-tasks in a laboratory experiment (less errors, less time needed) and increased favorable shopping behavior in a retail store (higher spending, faster product selection and more items chosen), compared to complex scents based on several ingredients (e.g., lemon and basil). Complex scents evoked the same responses as no scents and therefore provided no managerial benefit at all (Herrmann et al., 2013). However, Grabenhorst and colleagues (2011) show, in a functional magnetic resonance imaging (fMRI) study, that complex scents – including pleasant as well as unpleasant components – evoke an *attentional capture effect* that can influence internal and behavioral responses, while non-complex scents do not. Furthermore, in real-life situations, most scents consist of many components and, more importantly, mono-scents are said to be unable to evoke 'natural' (non-artificial) human responses (Kirk-Smith & Booth, 1987; von Kempski, 2002). Thus, we conclude that research has not finally clarified which degree of scent complexity is optimal for the context of ambient scents in a service environment.

Depending on the exposure duration, a physiological effect called *adaptation* might occur. Adaptation is a stimulus-induced sensitivity reduction, where a continuous exposure to a specific scent leads to a decrease in the perceived intensity. In most cases, such olfactory cell fatigue occurs within 10 seconds, depending on a scent's structure and concentration. After removing the stimulus, the olfactory cells' sensitivity usually recovers fully (Köster & de Wijk, 1991). In contrast, frequent exposure to a scent might lead to the effect of *habituation*, an experience-based sensitivity reduction. Habituation can lead to a decrease of the perceived intensity to an entirely unconscious perception after repeated contact with a specific scent (e.g., the smell of one's home) (Henshaw, 2014). The specific smell perception decreases over time, as the scent is rated as not significant by the respective brain areas and can thus be ignored (Engen, 1991; Poellinger et al., 2001). It must be noted that adaptation and habituation do not influence a scent's de facto intensity, but only the individually perceived intensity decreases.

Insight 4: Scents do **not** need to be **consciously perceived** to evoke reactions within exposed individuals.

Human scent perception and the formation of an actual smell impression are embedded in a complex two-step process of physiological and neurological procedures. The first step relates to the reception of an olfactory molecule by the physiological scent receptors in the nasal mucous membrane. The second step comprises multiple neurological and psychological processings of the perceived signals in various areas of the human brain. The actual perception of a scent can be processed consciously or unconsciously, while the potential of the scent to actually impact an individual is independent of the consciousness level of exposure (Binder et al., 2009; Doty, 2001; Pyrski & Zufall, 2009; Slotnick & Weiler, 2009). Thus, even if customers or employees do not perceive a (pleasant or unpleasant) scent, it might still (positively or negatively) affect the exposed individuals within a servicescape (Lee & Schwarz, 2012; Levy et al., 1997; Li et al., 2007; Lorig et al., 1991).

Insight 5: The perceived scent intensity depends on an individual's **scent acuity**, which is determined by numerous individual moderators such as demographical, physiological, situational, personality-related or experience-related factors (see the effect of habituation).

Olfactory acuity generally improves until one's 20s, remains stable between 20 and 60 years, and decreases after 60 (Doty, 1991a). Regarding the role of gender, women are said to generally have higher olfactory acuity than men (Brand & Millot, 2001; Chen & Dalton, 2005; Doty et al., 1985). Keller et al. (2012) show that young, non-smoking females have the highest average scent acuity.

Health status and pathological dysfunction of the olfactory sense can reduce or change olfactory acuity or can lead to a total loss of sensitivity within affected individuals (anosmia) (Strous & Shoenfeld, 2006; Tafalla, 2013).[10] Also, specific personality traits are said to influence olfactory acuity, such as being anxious or neurotic increases scent acuity while extraversion decreases it (Chen & Dalton, 2005; La Buissonnière-Ariza et al., 2013; Larsson et al., 2000; Pause et al., 1998).

Furthermore, situational or temporal factors can influence scent perception: Tiredness, reduced attention, and smoking decrease olfactory acuity (Frye et al., 1990; Vennemann et al., 2008), while the moderate consumption of alcohol (Engen et al., 1975) or cocaine (Douek, 1974) tends to increase it. One's emotional state also seems to influence scent acuity (Chen & Dalton, 2005). During pregnancy, women's scent acuity usually increases and often leads to avoidance of specific scents, such as cigarette smoke (Schenker, 2001).

The latest results even indicate that there are differences between cultures in general (Croy et al., 2014), especially between industrialized and non-industrialized populations. Possible explanations for indigenous people's higher sensitivity to scents might be less exposure to air pollution, better training owing to the higher importance of the sense of smell (Sorokowska et al., 2013), or genetic predisposition (Mainland et al., 2014).

Overall, service managers should be aware of an ambient scenting's target group, as the intensity must be adapted for their specific expected scent acuity. For instance, the intensity of scent diffusion in a retirement home should be considerably higher than in a young ladies gym if the intention is to reach a conscious perception by exposed individuals.

Insight 6: When consciously perceived, individuals can evaluate the **subjective characteristics** of a scent: its quality, hedonic value (preferences), and its perceived congruence.

A scent's *quality* relates to the verbal description of a smell based on previous experiences and evoked impressions using adjectives (e.g., sweet or rotten) or nouns describing a scent's supposed source (e.g., it smells like roses) (Pyrski &

---

[10] Interestingly, Keller et al. (2012) found evidence that decreased scent acuity influences nutrition (food choice and intake), thereby leading one to being over- or under-weight.

Zufall, 2009). "Certainly, odour words are virtually always names of the typical source object" (Kirk-Smith & Booth, 1987, p. 160). Furthermore, a scent's quality description might vary depending on its intensity (Takagi, 1989): For instance, in low concentrations, the substance *β-Ionon* smells like violets, whereas in higher intensity it smells like cedar (Legrum, 2011). Overall, the human nose is able to distinguish up to 10,000 scent qualities (Buck & Axel, 1991). However the science of identifying a scent's quality through its molecule structure is still in its infancy (Pyrski & Zufall, 2009).

Hedonic value relates to the emotional evaluation of a smell as either pleasant or unpleasant. Preferences for certain scents are very subjective and depend on learning and education (Ayabe-Kanamura et al., 1998; Gulas & Bloch, 1995). Scents are also directly linked with past experiences and events, which usually determine whether a smell is evaluated as pleasant or unpleasant. Therefore, almost every scent is connected with certain associations and provides an identity to the context in which we perceive it. This link is strengthened with the frequency of perception in a specific context (Orth & Bourrain, 2008; Sandoz, 2009). Studies with newborns show that without any previous experience, milky and fruity scents are rated as pleasant, whereas fishy odors and the smell of rotten eggs are evaluated as unpleasant (Doty, 1991b). A smell universally perceived as pleasant seems to be the scent of newborns (Lundström et al., 2013). Furthermore, the hedonic evaluation is interrelated with a scent's intensity (Takagi, 1989): Even a pleasurable fragrance can be perceived as unpleasant if the concentration is too high, and an unpleasant scent might not be evaluated as unpleasant if its intensity is low enough (Henion, 1971; Legrum, 2011; Winneke et al., 1995).[11]

We also try to further articulate the phenomenon of *congruence* in the context of scent perception. Schema theory defines congruence as a positive comparison between the perceived attributes of a service and an already existing schema of past experiences in a person's mind (Mandler, 1982). The crucial question is in which schema does a scent need to fit in the context of a specific service, that is, will it be congruent with the brand, the servicescape, or the managerial inten-

---

[11] For instance, concentrated phenylacetic acid stinks like stale urine from horses, while when greatly diluted it smells like honeycomb (Legrum, 2011).

tion the ambient scent seeks to support (e.g., creating a relaxing atmosphere)? Ideally, all three should align, for instance, the Four Seasons Hotel brand stands for exclusivity and luxury. Accordingly, its servicescape is elegantly designed, and smells like exquisite food, a spa, and cleanliness. Therefore, a pleasant ambient scent would be congruent with all the above-mentioned targets and various schemas. In the case of most public transportation companies (e.g., a railway company), however, congruence is harder to determine. Most have a poor brand image, and their servicescapes are usually old and/or dirty (e.g., train compartments). A pleasant ambient scent that should support an organization's goal to provide a pleasant transportation experience would therefore fit neither the brand image nor the servicescape. These examples illustrate the problem of scent congruence in the context of service management.

Olahut (2013) therefore argues that congruence should be considered as a separate scent dimension, and not just a moderator as Gulas and Bloch's (1995) model suggests. However, she too does not specify to what schema a scent should fit. From a research perspective, most studies choose their specific research object as reference point for the congruence evaluation (e.g., Bosmans, 2006; Mattila & Wirtz, 2001; Morrin & Chebat, 2005); for marketers, this solution is impractical. The problem of congruence is becoming even more complex, because one scent can never fit all, since each person has different olfactory experiences, brand images, or schemas activated in relation to a specific servicescape. Every scent can induce different smell perceptions, both within an individual (owing to situational or temporal moderators) as well as between individuals (owing to individual moderators), which makes it virtually impossible to develop and deploy 'the one and only' congruent scent.

Thus, it can be assumed that an ambient scent should at least not be incongruent – neither to the servicescape, nor the brand positioning, nor the scenting's managerial objective. Furthermore, we argue that it is likely that individuals will learn to adapt their individual schema accordingly when they repeatedly perceive a particular scent in the context of a specific service (Mandler, 1982).

Insight 7: Other **external moderators,** such as thermal conditions, and other stimuli present in the service environment influence ambient scents' perception.

First, the ambient scent and perceptions of it might vary depending on thermal factors, such as temperature, humidity, and air pressure in the indoor environment (Henshaw, 2014; Knoblich et al., 2003; Schön & Hübner, 1996; von Kempski, 2002). For instance, with increased room temperature different molecular components of a scent source volatize and individuals thus get a different smell impression.

Second, ambient scents' effects might be moderated by other stimuli present in the service environment (Gulas & Bloch, 1995). Individuals perceive their environment with all their senses (Ackerman, 1990; Krishna, 2012; Mehrabian & Russell, 1974), and if the respective visual, acoustic, haptic, and olfactory impressions are congruent or do not differ widely, general information processing is improved (Mandler, 1982).

The most prominent of our senses is vision (Ackerman, 1990): If what we see and what we smell fits, individuals are able to identify a scent faster and more accurately. Without a visual counterpart, the human brain is hardly able to identify or allocate a scent's source (Gottfried & Dolan, 2003). Similar results were achieved by Mani (1999), who showed that individuals are able to learn associations more accurately if a scent is congruent with visually presented colors or labels. Also, recent research revealed that if customers are confronted with a messy retail store, a pleasant ambient scent significantly worsens product evaluations, owing to a mismatch between unpleasant visual impressions and pleasant olfactory stimulation (Doucé et al., 2014).

Several studies have investigated the interaction between music and ambient scent: Congruent music seems to moderate ambient scent's effects; Spangenberg et al. (2005) revealed better evaluation of an environment and higher behavioral intentions in a setting with congruent acoustic and olfactory information cues. In line with this, Mattila and Wirtz (2001) could demonstrate that music and ambient scent that are congruent in terms of their arousal level lead to better evaluations of the environment, higher satisfaction, and more favorable shopping behavior in a mall.

Krishna et al. (2010) studied the interaction between sense of touch and sense of smell. The authors showed that if an ambient scent is congruent with haptic characteristics of a product (its texture and temperature), product evaluations significantly improved in a laboratory experiment.

However, too much stimulation might also lead to negative effects (Morrin & Chebat, 2005), as most individuals prefer intermediate levels of stimulation (Steenkamp & Baumgartner, 1992). Thus, managers should not forget that ambient scent is one aspect among many in overall marketing strategy and should be integrated cautiously and properly (Ravn, 2007).

Insight 8: Perceived **air quality** (PAQ) mediates ambient scents' influences on internal and behavioral reactions of exposed individuals, independent of the consciousness level of scent perception.

Bitner introduced air quality as an ambient condition in the servicescape model. We broaden this view as follows: *Perceived air quality* describes an individual's subjective perception of indoor air quality, and is considered crucial for the well-being and health of exposed individuals (Frontczak et al., 2012; Frontczak & Wargocki, 2011). Von Kempski (2002) suggests that indoor air must be objectively unpolluted (i.e., hygienically clean) as well as perceived as subjectively pleasant, natural, and fresh. From an olfactory perspective, the holistic environmental impression – the individually perceived servicescape (Bitner, 1992) – is equivalent to PAQ within the servicescape and is influenced by internal and external scent sources, and if applied also by diffused ambient scents. Hence, even if no ambient scents are deliberately introduced into a servicescape, the olfactory situation in the environment will still influence individuals through their perceived air quality. If a service provider decides to use pleasant ambient scents, its introduction should lead to improved PAQ (von Kempski, 2002, 2004). In turn, the improved air quality should lead to an indirect positive affectation of exposed individuals. We therefore consider PAQ an important mediator variable in the context of scent research.

Insight 9: Olfactory stimulation evokes **various reactions in humans**, ranging from mere physiological responses, to emotional reactions, and to complex cognitive responses.

Gulas and Bloch (1995) only consider affective responses on customers in their model. However, as Bitner (1992) suggests, scents can also induce physiological and cognitive reactions in both customers and employees, as they are part of the same environment: For starters, olfactory stimulation can induce physiological activation (activation of the CNS), and can therefore trigger an alteration of skin conductance responses (Møller & Dijksterhuis, 2003), human brain activity (e.g., Masago et al., 2000; Owen & Patterson, 2002; van Toller et al., 1993), blood pressure (Heuberger et al., 2001; Heuberger et al., 2004; Höferl et al., 2006), and dilation of the pupils (Schneider et al., 2009). Owing to the activation of the CNS, olfactory stimulation can enhance the processing and storage of scent information as well as of other stimuli perceived in the same context, such as sounds, textures, etc. (Gould & Martin, 2001; Rempel, 2006). This seems promising for marketers, because scents could be generally used to activate exposed subjects in a servicescape.

Olfactory stimuli can affect bodily functions during sleep (Badia et al., 1990): Some scents enhance sleeping efficiency and reduce movement while sleeping (Goel et al., 2005; Raudenbush et al., 2003). This could be interesting for hotels in order to improve their customers' sleep quality. Furthermore, specific scents can reduce the perceived intensity of pain (Marchand & Arsenault, 2002; Villemure et al., 2003). In a marketing context, this could be relevant, for instance, to reduce perceived pain in piercing or waxing studios.

Scents can also induce emotional reactions in customers and employees. On the one hand, several studies show that the introduction of a pleasurable ambient scent can positively affect individuals' emotions (Mattila & Wirtz, 2001; Michon et al., 2005; Morrison et al., 2011; Schifferstein et al., 2011; Spangenberg et al., 2005), defined as an internal state of readiness that is mostly intentional (Bagozzi et al., 1999); as well as mood (Baron, 1990; Baron & Thomley, 1994; Knasko, 1995; Warm et al., 1991), which is fairly diffused, lasts longer, and has a lower intensity than emotions (Bagozzi et al., 1999). On the other hand, several other studies are not able to reveal any effects of scent on people's mood or emotions (Gilbert et al., 1997; Knasko, 1993; Ludvigson & Rottman, 1989; Mitchell et al., 1995; Morrin & Chebat, 2005; Teller & Dennis, 2012). One explanation for these inconsistent results might be that different ambient scents were used – despite all being rated as pleasant by the respondents – which were linked with different previous personal experiences, and thus led to different effects on affective responses. Overall, it seems certain that unpleasant scents worsen mood and lead to negative emotions (Asmus & Bell, 1999; Knasko, 1992; Rotton, 1983). Therefore, avoiding unpleasant odors within a servicescape should be a top priority for service managers.

Regarding memories, there is evidence that even after a long time, events can be remembered better if they are stored in combination with a specific scent (Baeyens et al., 1996; Epple & Herz, 1999; Herz & Cupchik, 1992; Laird, 1935; Lehrner et al., 2000; Lehrner et al., 2005; Robin et al., 1999). This effect is known as the *Proust phenomenon* (Herz et al., 2004; Miles & Berntsen, 2011) and relates to Marcel Proust's novel *Swann's Way (1913)*, where the main character perceives the smell of freshly baked *Madeleines*, which unleash lively and very emotional childhood memories. This memory power of scents might help service companies to link their offerings to a specific ambient scent, thereby increasing the recall and recognition potential of their services, even after a long time.

Also, there is evidence of an influence of ambient scents on customers' cognitive responses. The presence of a pleasant ambient scents can induce customers to evaluate the (service) environment (Michon et al., 2005; Morrin & Chebat, 2005; Parsons, 2009; Schifferstein et al., 2011; Spangenberg et al., 1996; Spangenberg et al., 2005; Spangenberg et al., 2006), as well as present service employees (Baron, 1981, 1983, 1986; Maille, 2006; Mattila & Wirtz, 2001) more positively. Further studies demonstrate an impact of ambient scents on the evaluation of service quality (Girard et al., 2013; Maille, 2006; McDonnell, 2007; Morrin & Chebat, 2005), service experience (Girard et al., 2013), or a brand (Girard et al., 2013; Mani, 1999; Morrin & Ratneshwar, 2000, 2003). Olfactory stimulation also seems to affect sense of time: Subjects evaluate the perceived dwelling, waiting, and shopping time as shorter in a scented environment (Maille, 2006; Spangenberg et al., 1996; Spangenberg et al., 2006). Even, it seems possible to positively influence overall customer satisfaction by means of ambient scents (Mattila & Wirtz, 2001; Morrison et al., 2011).

Regarding scents' potential to influence employees' cognitions, we could identify several studies that conducted laboratory experiments in work-related contexts. The results suggest that ambient scents can alter the evaluation of the (work) environment (Baron, 1990; Baron & Bronfen, 1994), other persons present (Baron, 1981, 1983, 1986; Fiore & Kim, 1997; Li et al., 2007; McGlone et al., 2013), an individual's self-evaluation in terms of goal and performance assessments (Baron, 1990; Gilbert et al., 1997; Knasko, 1992, 1993), and positively influence employees' perceived stress levels (Baron & Bronfen, 1994; Baron & Thomley, 1994).

As a result, service managers might use ambient scents as an instrument to positively influence various physiological, emotional, and cognitive reactions in both customers and employees present in a servicescape.

Insight 10: Scents can induce approach or avoidance **behavior** in customers and employees in or towards a servicescape.

Ambient scents seem to raise exploration and information search tendencies in customers (Doucé et al., 2013; Mattila & Wirtz, 2001), the willingness to buy and pay (Fiore et al., 2000; Herrmann et al., 2013), as well as the intention to revisit (Spangenberg et al., 1996; Spangenberg et al., 2005; Spangenberg et al., 2006). Besides scents' effects on perceived dwelling time, there is also evidence for such an influence on the actual length of stay in an environment (Guéguen & Petr, 2006; Maille, 2006). Scents can even enhance behavior, for instance, dancing in a nightclub (Schifferstein et al., 2011), the conversion rate from store visitors to paying customers (Jacob et al., 2014), as well as the amount of items bought (Spangenberg et al., 2006), and shoppers' overall expenditures (Chebat et al., 2009; Guéguen & Petr, 2006; Hirsch, 1995; Jacob et al., 2014; Morrin & Chebat, 2005; Spangenberg et al., 2006).

Concerning possible effects on employees, experiments on work-related behavior show that some olfactory stimuli can enhance individuals' physical performance in routine jobs (Barker et al., 2003; Ho & Spence, 2005; Raudenbush et al., 2002; Sakamoto et al., 2005; Warm et al., 1991), mathematical operations (Degel & Köster, 1999; Gilbert et al., 1997; Knasko, 1993; Ludvigson & Rottman, 1989), and linguistic tasks (Baron & Bronfen, 1994; Baron & Thomley, 1994; Degel & Köster, 1999; Herrmann et al., 2013; Knasko, 1993; Rotton, 1983), whereas unpleasant scents divert employees and lead to less responsiveness and performance, as well as higher error rates (Habel et al., 2007; Nordin et al., 2013; Rotton, 1983). Furthermore, unpleasant scents increase employees' motivation to escape the environment (Asmus & Bell, 1999). To date, there seems to be no evidence for any impact of olfactory stimulation on creativity tasks (Degel & Köster, 1999; Knasko, 1992).

Besides approach and avoidance behavior, other reactions to scents also seem possible. In Bitner's servicescape model (1992), individual behavior evoked by a service environment mainly includes physical behavior within the environment, defined as "physical movement toward, or away from, an environment or stimulus, degree of attention, exploration, favorable attitudes such as verbally or nonverbally expressed preference or liking, approach to a task (...), and approach to another person" (Mehrabian & Russell, 1974, p. 96). However, other reactions to a servicescape seem possible, such as staying within an unfavorable situation, complaining, or asking for help (Hirschman, 1970). To expand the original servicescape model beyond approach and avoidance behavior, we intend to integrate the concept of coping.

The concept of coping stems from stress research and implies "constantly changing cognitive and behavioral efforts to manage specific external and/or internal demands that are appraised as taxing or exceeding the resources of the person" (Lazarus & Folkman, 1984, p. 141). Thus, coping has two main functions: The cognitive regulation of emotional responses caused by stress (emotion-focused coping)[12] as well as actions taken to change the situation that is perceived as being stressful (problem-focused coping) (Folkman et al., 1986). Besides escaping from an unpleasant environment, which is similar to avoidance behavior, Aldwin and Revenson (1987) identify other possible behavioral reactions: To passively remain within the situation when alternatives would harm more than improve the situation (exercised caution); to actively seek to change the unpleasant situation (instrumental action); to request assistance from third parties (support mobilization); or to seek to express displeasure and/or to get something positive from the situation through negotiation, bargaining, or compromise (negotiation). Scent-related examples in service settings might be: During a train ride, avoidance is hardly possible, but one can change compartments or hold one's nose. When sitting near a smelly toilet in a restaurant, one can ask for another table or advice the service employees to clean the

---

[12] In this context, stress is defined as "unfavorable person-environment relationship" (Lazarus, 1993, p. 8).

toilet. From an employee's perspective, avoiding the workplace is not realistic, but for instance if it stinks in the office, one might change the situation by opening a window or actively scenting the environment with a pleasurable perfume.

In our scentscape model, we integrate these possibilities by adding behavioral coping as additional response to complement approach and avoidance behavioral.[13,14]

Insight 11: Scents can affect the quality and quantity of **social interactions** between customers and employees in the service encounter as well as among each other.

Few studies have investigated the influence of ambient scents on the quantity and quality of social encounters (Shostack, 1985) – a remarkable research deficit, since social interaction is often a crucial aspect, especially in services (Zemke & Shoemaker, 2007). The identified studies imply that pleasant olfactory stimulation can have a positive impact on the possibility and frequency of social interactions (Doucé et al., 2013; Zemke & Shoemaker, 2007, 2008). Scents might also improve interaction quality, as they can lead to enhanced bargaining power (Baron, 1990), which might be interesting for sales pitches or conferences. Scents might also be able to enhance people's willingness to help (Baron, 1997; Baron & Thomley, 1994), and lowers conflict potential of exposed individuals (Baron, 1990). These aspects might be relevant to improve safety precautions, for instance, at mass events, such as rock concerts or ice hockey games.

Taken together, the outlined scentscape model provides a rich integrative framework that explains the perception and effects of scents in a service context. We could show that one and the same scent might smell differently to different people and might trigger a variety of different (physiological, emotional, cognitive, and behavioral) reactions. The managerial challenge of controlling a systematic influence of scents in a servicescape becomes clear.

---

[13] We do not discuss emotion-focused coping separately, because it can be subsumed under internal (emotional and/or cognitive) reactions, and is therefore already covered by Bitner's servicescape model. It must be noted that, in stressful situations, the two forms of coping usually occur together (Folkman et al., 1986).

[14] Rigorously, one could subsume avoidance behavior as a sub-aspect of coping. However, we decided to leave it as a separate category, since approach and avoidance behavior are well-established expressions since they were introduced by Mehrabian and Russell in 1974.

From a research perspective, we recognize that there might be further interaction effects between some of the factors introduced in the scentscape model. However, a fine-grained understanding of such interactions would require an in-depth research program, which is beyond the scope of this article.

4   Discussion and Implications

The integration of ambient scents is a challenge for companies that recognize the importance of the olfactory situation within a servicescape. We have advanced a new framework, which outlines that – depending on the service type – scent-specific characteristics must be considered, and that scents will lead to non-observable internal processes of scent perception as well as observable reactions in both customers and employees. The ideas put forward in this article lead to a number of managerial as well as research-related implications.

4.1   Managerial Implications

Managers and marketers increasingly utilize ambient scents within service organizations' physical surroundings (Elejalde-Ruiz, 2014), but unfortunately the ultimate impacts of such olfactory stimuli on individuals in a service facility is not yet fully understood. Several challenging managerial implications arise when considering the scentscape's influence on customers and employees. Overall, a careful yet creative management of the olfactory situation in a scentscape can contribute to the achievement of a company's external marketing goals as well as its internal organizational goals. The scentscape provides a visual framework that guides managers on relevant issues and questions that should be raised and answered before introducing ambient scents into a service environment (see Table 2).

Table 2: Questions Practitioners Should Address[15]

Service type
What is the service type of your business? The stronger the agreement with the following questions, the more relevant ambient scents are:
- Is it a service with a high extent of explanatory aspects, where the understanding of additional information is crucial?
- Is it a service where the actual service experience is a key aspect of your offering?
- Is it a service that is hard to evaluate owing to a high extent of credence qualities?
Prevailing olfactory situation
- What is the prevailing olfactory situation in the servicescape?
- Which already present internal and external scent sources must be considered?
- To what extent can these other scent sources be controlled?
Scent characteristics
- What arousal level is appropriate for your customers and for your employees?
- Will the exposed individuals generally perceive the scent consciously or unconsciously?
- Is the scent considered as not unpleasant by most of the target groups?
- Which reference object of scent congruence matters most for your company?
- Is the scent at least not considered as incongruent with your servicescape, your brand's positioning, and the intention of introducing a scent (e.g., creating a relaxing experience)?
- How long will individuals be exposed to the ambient scent: Do adaptation effects need to be considered?
- How frequently will individuals be exposed to the ambient scent: Do habituation effects need to be considered?
Internal and behavioral responses
- What are desired responses in customers and employees?
- What are the long-term effects on the exposed individuals?
Additional issues
- How one deals with concerns of manipulation: Will ambient scent use be made transparent (e.g., by signs at the entrance, *This room is scented for your convenience*)?
- Will ambient scent use be actively promoted in order for the organization to be perceived as innovative?

Ambient scent constitutes an additional environmental cue. Its importance depends mainly on the service typ. If the service holds a high extent of credence qualities or a high extent of explanatory aspects, ambient scents can act as sensory information and might facilitate the evaluation of a service and its quality (Meyer, 1991; Zeithaml, 1981): For instance, in a hospital, ambient scent can enhance the perceived cleanliness, and patients might take this as an indication of high-quality medical treatment. But also, for settings where the service experience is key, for instance in nightclubs, ambient scent can act as an additional sensory experience cue and can enrich the environment (Berry et al., 2006). Arousing scents can also support the processing of other available cues and information (Gould & Martin, 2001), which generally might help customers to evaluate a service.

---

[15] Own illustration.

Even if an organization does not deliberately introduce ambient scents into its servicescape, the olfactory situation in a service environment still influences present customers and employees via their perceived air quality (von Kempski, 2002); for instance, if many people perspire in a fashion store on a hot summer's day, the PAQ would be poor. Such external and internal factors are mostly hard to control; therefore, active indoor air quality management is critical to ensure pleasant and fresh air, so as to avoid a potential negative effect of unpleasant olfactory experiences in the service delivery process.

If managers decide to introduce ambient scents in their servicescape, they should not see scents as one-dimensional stimuli. The selection of an olfactory stimulus depends on the intention sought through scent use: If the objective is to create a relaxing atmosphere in a spa, the arousal potential should be low, whereas in a fitness center, an energetic scent may be chosen. Furthermore, a scent's quality and hedonic evaluation interact with its intensity – even a pleasant scent can become irritating if its intensity is too high. The right intensity also depends on the target group's scent acuity. At a retirement home, for instance, the stimulus' intensity should be higher than in a youth-oriented fashion store.

In general, when thinking about introducing ambient scent into a servicescape, it should be perceived as pleasant by the target group, bearing in mind that almost no scent is universally evaluated as pleasant. Therefore, we recommend that a scent should be rated at least as not unpleasant (i.e., neutral or pleasant) by the majority of the target groups (customers and employees). The scent preferences of the managers choosing a scent are secondary.

A further challenge lies with a scent's congruence. As noted, there are multiple reference objects, and a scent should at least not be incongruent with the brand's positioning, the servicescape (including other stimuli present, such as music, lightning, and design), and the scenting intention (e.g., creating a relaxing atmosphere). For instance, the fashion retailer Abercrombie & Fitch creates a multisensory experience environment in which everything fits together: The brand image is young and wild, and so is their servicescape – created like a disco, with loud music, dancing employees, as well as gritty lightning and atmosphere. The ambient scent, labeled *Fierce*, is intense and wild. However, the concept always needs to fit to the specific target customer group, as most elderly people would probably feel overwhelmed by the stimuli overload in Abercrombie & Fitch's servicescapes (Steenkamp & Baumgartner, 1992). Therefore, managers should carefully coordinate and integrate congruent sensory stimuli within a service environment rather than to randomly combine environmental cues.

Depending on the service type, odor adaptation should be considered, for employees in general, and also for customers remaining in a servicescape for a while (e.g., going to the movies). In contrast, for frequent visitors of a servicescape, or employees repeatedly exposed to olfactory stimulation in their workplace, the effect of habituation is more likely. Owing to these, ad hoc observation of scent reactions could be insufficient, as customers and employees might no longer cautiously perceive the scent, but it could still exert an influence via unconscious perception. Thus, managers should measure and monitor ambient scent's impacts in the long run, to cautiously and consistently assess managerial benefits.

In general, "predicting specific odor effects (i.e., specific moods, thoughts, attitudes, or behaviors) [is] a risky business" (Fitzgerald Bone & Scholder Ellen, 1999, p. 259). Thus, ambient scent's influences on customers and employees should always first be tested in a pilot. A specific ambient scent can trigger various physiological, emotional, cognitive, and behavioral responses. However, the desired reactions in customers are not necessarily the same as for employees. For instance, lavender, which is relaxing, might be ideal for a customer's relaxation in a spa, but not for the employees' productivity, as they need to be aroused in order to consistently deliver excellent service. Furthermore, not only positive reactions are to be expected; for instance, a very intense scent in a fashion store might work for a young target group that seeks a disco-like shopping experience, but among the employees present in the store for eight hours a day, it might lead to behavioral coping or avoidance (Asmus & Bell, 1999; van Harreveld, 2001). Managers should therefore balance and align their targets regarding desired customer responses to pleasurable ambient scents with their employees' interests (or vice versa).

Because scent perception can be conscious and unconscious, scents could be considered as a manipulative marketing attempt used by an organization to inappropriately influence its customers and/or employees. Owing to the olfactory sense's specific characteristics, ethical considerations should be taken into account (Bradford & Desrochers, 2009; Lunardo, 2012; Lunardo & Mbengue, 2013). As people smell with every breath they take and breathing is vital for life, it is hardly possible to fully elude olfactory stimulation (Ackerman, 1990). This is why it seems appropriate to make transparent use of ambient scents in a servicescape. Customers and employees should at least be informed about the application of scents and, optimally, they should be able to choose whether or not they want to be exposed to these scents (e.g., trough scented and unscented hotel rooms). Otherwise, reactance[16] cannot be precluded, and the expected positive effect of a pleasant scent could be reversed by negative emotions towards a company (Bradford & Desrochers, 2009; Lunardo, 2012). However, by openly communicating the use of ambient scent (e.g., via signs at entrances), the scentscape can support the signaling of an organization's positioning and can provide a stronger distinction over its competitors.

In sum, the scentscape provides an overview of a large number of crucial factors that need to be considered by managers when introducing ambient scents, but also on possible consequences that might occur through pleasant or unpleasant air quality. Hence, we strongly recommend that service managers acknowledge ambient scents and the scentscape model as a manageable element of the organization's environment and marketing strategy.

4.2 Research Implications

The integrative scentscape framework also proposes a variety of interesting research implications. Table 3 sets out key questions for further research, organized around the elements of our conceptualization.

---

[16] The essence of the theory of psychological reactance is that individuals will resist attempts to manipulate them (e.g., ambient scent use) in given circumstances. Psychological reactance is a motivational state focused on the restitution of an individual's freedom. For instance, if customers feel manipulated or betrayed by ambient scent usage, subjective reactance (e.g., negative word-of-mouth) as well as behavioral reactance (e.g., a change in service provider) may occur (Brehm, 1966; Wicklund, 1974).

Table 3: Questions for Further Research[17]

Service type
- Are there any differences in the impacts of ambient scent between service industries?

Scent characteristics
- What are the long-term effects of ambient scent after extended or repeated exposure?
- What happens after the ambient scent use is discontinued after a period?
- Is it enough for an ambient scent to be perceived as pleasant?
- Which reference object of scent congruence is most important to achieve positive results (e.g., the brand positioning, the servicescape, etc.)?

Internal and behavioral responses
- How does ambient scent affect employee motivation, satisfaction, stress, coping ...?
- How does ambient scent affect customer service value, perceived risk, stress, coping ...?
- How does ambient scent affect the social interaction between customers and employees, as well as among each other?

Mediation PAQ
- Does perceived air quality mediate the influence of ambient scent on individual responses?

Individual and external moderators
- How do different combinations of individual characteristics affect scent acuity and influence?
- How do ambient scents interact with other stimuli present in the servicescape?

The scentscape is intended as a trigger for future empirical research to examine individual scent variables and their possible influences in the context of varied service organizations. The striking lack of research regarding the investigation of scent effects in service businesses generally, with the exception of the retail sector, also calls for research on the different importance and differential effects of the scentscape across various service industry types, for instance, depending on the extent of inherent credence and experience qualities. Our model therefore constitutes a first theoretical step towards closing the gap between the topic's high practical relevance and the prevailing lack of service research.

The scentscape provides a visual overview of a range of influencing factors and consequences that should be considered when investigating ambient scent's effects. In most empirical research, scent is seen as a one-dimensional stimulus, whereas in fact, the interaction of the different objective and subjective characteristics should be considered and further investigated. Regarding the time dimension, to date, no long-term studies have investigated the effects of repeated or enduring scent exposure in a marketing context. There is both a theoretical and practical need to explore whether beneficial short-term influences of scent exposure shown in ad hoc studies might, over time, level out or even turn negative. Based on optimal arousal theory (Berlyne, 1971; McClelland et al., 1976), we assume that a novel ambient scent stimulus has a stronger

---

[17] Own illustration.

influence when it is first introduced than in the long run, but this assumption still needs further research. Furthermore, an interesting question relates to possible aftereffects: What happens if permanent scent diffusion is removed after a while? Will individuals miss the pleasant ambient cue?

Further questions for research include the hedonic evaluation and the congruence of a scent. Is it enough for a scent to be rated as pleasant to evoke positive effects, and if so, how positive should the hedonic evaluation be? Is it sufficient for a scent not rated as unpleasant? Regarding scent congruence, which reference object the scent should fit to is the most important one – the brand, the servicescape, or the managerial intention of scent use?

Our literature review illustrates the necessity of increasing research on the effects of ambient scents on employees. To our best knowledge, to date, there are only a few laboratory experiments in work-related contexts. Therefore, we call for further investigation in this underrepresented research area; for instance, the examination of a scent's impacts on emotional and cognitive internal responses such as motivation, job satisfaction, and commitment of employees present in a scented servicescape. Research may also consider a scent's influence on customers, of which some aspects still need to be addressed in future studies, including scents' influence on perceived service value or risk. The newly integrated concept of behavioral coping should also be investigated for both customers and employees. Also, few studies have focused on the impacts of ambient scents on social interactions. Because the social encounter during the 'moment of truth' is a crucial aspect in the service delivery process of many services (Bitner et al., 1994), it becomes clear that further research regarding ambient scents' influence on social interactions is inevitable and urgently needed.

Furthermore, despite an extensive literature review, we could find no empirical study in marketing or service research on the role of perceived air quality. The insights from air conditioning research suggest that PAQ could be a variable mediating ambient scent's influences on the internal and behavioral responses of exposed customers and employees. The scentscape calls for future investigation, to validate this relationship in an empirical service setting.

Also, a range of possible moderators needs further investigation. For instance, how do different combinations of individual characteristics affect scent acuity?

Or: How do ambient scents interact with other stimuli present in the servicescape? The last question is probably one of the most important aspects for managers, as ambient scents are only one aspect among many within the sensory appearance of service environments.

Most of the studies we found examine scents' influences via laboratory experiments. Following the discussion about rigor vs. relevance (Varadarajan, 2003), more field experiments on scents' influences with real brands as well as real customers and employees would be beneficial to validate previous findings in real-life situations. Furthermore, in general, experimental methods and surveys are considered appropriate for assessing the scentscape's impacts. In marketing and services research, the use of latent constructs for indirect measurement techniques of dependent variables is common. However, we suggest that direct measurements from neuroscience (e.g., via fMRI) could also be beneficial in marketing scent research, since olfactory stimuli's influences on the different cerebral areas could be studied more directly and in greater depth.

In short, the scentscape constitutes an attempt to increase existing theory regarding ambient scents' effects in services research. The framework provides a wide range of directions for future research on the stimulus scent itself, on dependent variables, and on adequate research methodologies: "Olfaction has always been seen as the least important of our senses, but I think that if science is able to devote it the time and energies it deserves, olfaction (…) could be a key that will open many important aspects of our nature" (Tafalla, 2013, p. 1296).

5   Conclusion

The use of ambient scents in service practice is widespread. Unfortunately, most companies use olfactory stimuli without truly understanding the scent perception process and its effects on their customers and employees. Therefore, we developed the scentscape – an integrative framework that leverages insights from service and scent research to describe scents' influences in service environments on employees and customers (MacInnis, 2011). We sought to make a conceptual contribution to service literature by relating and integrating different disciplines and by providing an overview of relevant influencing factors and consequences of scent use for scholars and marketing managers. Therefore, we set out research questions for further investigations, as well as crucial facts to guide practitioners in integrating ambient scents within their marketing programs.

C. ARE YOU ON THE RIGHT SCENT? AMBIENT SCENTS' SHORT- AND LONG-TERM EFFECTS ON CUSTOMERS IN A SERVICESCAPE

ANNA L. GIRARD

0   Abstract

The systematic use of pleasurable ambient scents is an intensifying trend in service and retail companies. Interestingly, however, the development of using scents among practitioners has seen little attention in marketing research, especially concerning the long-term impacts of such practices. Therefore, we conducted a controlled field experiment over four months with a customer panel in a German service company. We applied a pretest/posttest control group design to reliably measure dependent variables over time. According to our study, repeated scent diffusion over an extended period positively impacts the exposed service customers. The systematic use of ambient scents can therefore be an effective instrument for companies seeking to improve customers' service experience, quality, and perceived value in the long term.

Keywords: ambient scent, servicescape, long-term effects, customer responses, field experiment, optimal arousal theory

Acknowledgments: The author acknowledges valuable comments by Prof. Dr. Ingo Weller, Prof. Dr. Marko Sarstedt, and Dr. Marcus Demmelmair. The author also thanks Dr. Marc Girard for his general feedback and support as second coder in qualitative content analysis. For their assistance during the study, the author thanks her seminar students: Maximilian Abele, Emily Falk, Franziska Ferdinand, Marina Fuchs, Julia Huhndorf, Zeno Lobe, and Caroline-Sophie Märtens.

1   Introduction

Retailers and service providers have used ambient scents in their services-capes for several years now, and this trend is becoming increasingly important to practitioners (Elejalde-Ruiz, 2014; Klara, 2012). For instance, the public transportation provider Singapore Airlines introduced its ambient scent *Stefan Floridian Waters* in the 1990s already, and has holistically integrated the scent into its corporate communication concept as employee perfume, ambient scent, and cosmetic articles on board all aircrafts. The fashion retailer Abercrombie & Fitch has used its brand scent *Fierce* for several years and applies it extensively in stores, textiles, and employee perfumes. The Starwood hotel group gave each of its sub-brands (Four Points, Westin, and Sheraton) a distinct fragrance, using it as ambient scent mostly in their hotel lobbies (Bell, 2007; Lindstrom, 2005a).

To get an impression, the global market for air fresheners in the private sector is forecast to reach US$8.3 billion by 2015 (Global Industry Analysts Inc., 2012). Compared to this, the market for ambient scents, at an estimated $200 million spend in 2013, is still emerging, with an annual growth rate of 10% and an optimistic forecast of even $500 million until 2016 (Elejalde-Ruiz, 2014; Vlahos, 2007). The managerial goals of companies seeking to leverage ambient scents in their physical environments are diverse: For instance, covering bad smells present at a service delivery location, creating a pleasant atmosphere, enhancing a brand's recall value by diffusing a memorable brand scent, or advancing a customer's well-being or service experience – with the ultimate objective of increasing sales (Goldkuhl & Styvén, 2007; Knoblich et al., 2003; Kroeber-Riel & Weinberg, 2003; Mücke & Lemmen, 2010).

Interestingly, however, the diverse use of ambient scents in practice is not appropriately supported by marketing research, as it "is still a rather neglected area in services marketing literature" (Goldkuhl & Styvén, 2007, p. 1300), especially concerning the de facto effects of service providers exposing customers to a repeated and thus long-term olfactory influence. This is surprising, since Homburg et al. (2006) suggested that marketing phenomena should also be investigated from a dynamic (long-term) perspective, and that "[i]gnoring the time-variant character of servicescape effects may lead to inappropriate conclusions" (Brüggen et al., 2011, p. 72). Even reviewing the literature in scent-related sciences, such as medicine or psychology, provides only limited insights: We could identify only two recent studies on repeated scent exposure's effects on humans (Delaunay-El Allam et al., 2010; Sakamoto et al., 2012). Nonetheless, companies justify ambient scent use with various empirical findings of a positive influence of olfactory stimulation on customers' affective, cognitive, and behavioral reactions in cross-sectional, static, one-time experiments (for a review see e.g., Girard et al., 2015; Olahut, 2013).

But what happens over time? What impacts does a permanent integration of ambient scents into a servicescape have on customers that frequently visit a specific servicescape? For instance, if customers' positive responses to an ambient scent wear off over time, assessing only short-term effects might lead to an overstatement of the overall impact as well as suboptimal marketing resource allocation. This raises another question: What happens if a company decides to discontinue scenting, for instance owing to necessary budget cuts? Do customers get used to a great-smelling environment and react differently (e.g., negatively) if the ambient scent is suddenly no longer offered?

Thus, we are convinced that both a theoretical as well as a managerial need exists for a more in-depth understanding of the short-term and long-term effects of ambient scents' use and possible aftereffects on customers in retail and service environments.

By addressing these questions, this paper seeks to contribute to the limited research in three ways. First, we want to contribute to marketing literature by reviewing existing literature on scents' long-term effects from various disciplines and transferring the findings into the marketing context. We also introduce optimal arousal theory in order to theoretically explain possible short-term effects, long-term influences, and aftereffects of ambient scent exposure on customers. Second, based on the conceptual work, we investigate the temporal structure of ambient scent impacts, namely short-term and long-term effects, in a controlled field experiment. For this purpose, we collect customer survey data in a public transportation service company, where a pleasurable ambient scent is diffused via the air conditioning system. The survey data repeatedly measures affects, cognitions, and behavioral intentions of the same regular customers (commuters) at nine points in time over a four-month period. In our research design, we also examine possible aftereffects once the scent is removed. Furthermore, we conduct qualitative interviews with participants after the end of the experiment in order to gather additional customer insights to further confirm and interpret our empirical findings. Thus, our empirical contribution will be relevant for service and retail companies that are considering permanently introducing an ambient scent into their respective servicescape. Third, we also seek to make a methodological contribution by conducting – to our best knowledge – the first field experiment based on a longitudinal repeated-measures design with a customer panel in scent marketing research.

To outline our article, we will first review related literature and then theoretically derive hypotheses for the empirical investigation. Next, we describe our field experiment, the scent stimulus, and data collection, then discuss our analysis and present our results. Finally, we draw conclusions, highlight relevant limitations, and discuss key managerial implications as well as areas for further research.

2   Hypotheses Development

To develop a basis for our experiment, we first discuss insights from related literature before deriving this article's hypotheses.

2.1   Related Literature

We reviewed studies from scent-related sciences such as psychology, medicine, and chemistry that investigate effects of a repeated indoor scent exposure (more than once, at least) in an experimental setting on humans.[18] In total, we could only identify two studies on long-term scent effects on individual behavior. Data on long-term emotional or cognitive scent effects is apparently not available.

Sakamoto et al. (2012) show in a longitudinal study with a continuous and repeated olfactory stimulation over 360 days, that elderly nursing home residents are less aroused and fall much less frequently in a lavender-scented experimental group than in the control group without scent. A possible explanation is

---

[18] We exclude studies on the effects of bodily odors, and the effects of smoking or illnesses.

lavender's potential positive effect on individuals' arousal level and sense of balance. In a marketing context, higher relaxation could be beneficial for many service settings, such as medical treatments, spas, and hotel stays. An improved sense of balance could make traveling by ship more comfortable, for instance.

Delaunay-El Allam et al. (2010) show that toddlers of seven months are significantly more likely to choose a chamomile-scented teething ring than the control group and to use it for a longer time if their mother used chamomile balm during breastfeeding.[19] After 21 months, the experimental group also preferred a chamomile-scented bottle compared to the control group. The choice, but not the de facto water consumption was significantly influenced by previous experiences with chamomile scent. Thus, a repeated scent exposure is able to influence behavior: It might lead to a higher probability of choosing a scented product or service and might induce more intense usage.

Overall, both presented studies indicate that a repeated exposure to ambient scent might influence individual behavior. Furthermore, Delaunay-El Allam et al. (2010) conclude that "odor experience influences cognitive and affective processes (...) and that olfaction is a suitable model system to explore the formation of long-term memory" (p. 850). Nonetheless, there are no empirical findings to prove these assumptions in general or in a marketing context. This paper seeks to reduce this research gap and investigates possible scent influences on affect, cognition, and behavior in a service setting.

2.2   Hypothesized Effects of Ambient Scents on Affects, Cognitions, and Behavioral Intentions

This paper focuses on addressing three main research questions, which we elaborate in more detail in the following:

RQ 1. What are the short-term effects of one-time exposure to ambient scents on customers in a servicescape?

---

[19] The exposure period to chamomile scent varied between 8 to 120 days with an average of 47 days. The frequency of exposure varied between 10 times a day directly after birth until once a week and lasted between 10 to 30 minutes (Delaunay-El Allam et al., 2010).

RQ 2. What are the long-term effects of repeated exposure to ambient scents on regular customers in a servicescape?

RQ 3. What are aftereffects on regular customers after the removal of a longitudinally diffused ambient scent in a servicescape?

2.2.1 Short-term Effects

The servicescape, defined as "the immediate physical and social environments surrounding a service experience, transaction or event" (Bitner, 2000, p. 48) can act as physical evidence, which supports customers in evolving their emotions, cognitions, and behavioral intentions towards a service (Baker et al., 2002; Brüggen et al., 2011). "As tangible evidence will be seen, heard, felt, tasted, or smelt by the customer, the five senses play an important role in tangibilisation of services" (Goldkuhl & Styvén, 2007, p. 1298).

In particular, ambient scents as environmental cues form an integral part of a service environment, and might thus influence how it is perceived. Influenced by individual moderators or situational factors, olfactory stimuli might cause internal reactions within customers (and employees) and might therefore affect their actual approach or avoidance behavior (Bitner, 1992, 2000). Various empirical studies support Bitner's assumptions, and document that pleasant ambient scents positively influence affective and cognitive responses (for a review see e.g., Girard et al., 2015; Olahut, 2013), such as emotions, perceived service quality, and service value (Girard et al., 2013; Morrin & Chebat, 2005; Spangenberg et al., 2005), as well as behavioral constructs, such as word-of-mouth and loyalty (Herrmann et al., 2013; Spangenberg et al., 1996). Mattila and Wirtz (2001), for instance, showed that a minimal modification of the sensory environment, such as the diffusion of a pleasant ambient scent, led to a more positive evaluation of a store environment. Even though these studies' settings were cross-sectional and static, their results seem particularly applicable shortly after the introduction of an ambient scent, when customers first face a new environmental cue (Baker et al., 2002; Brüggen et al., 2011). Hence, we refer to them as short-term investigations.

Optimal arousal theory provides a theoretical explanation, postulating that every environmental stimulus has an optimal arousal level. The so-called *adaptation level* (AL) refers to the expected stimulus, which is determined by previous sen-

sory and perceptual experiences. A discrepancy between the de facto perception and its adaptation level (as the expectation of an event) causes a response
on the part of the individual (McClelland et al., 1976): "Certain stimuli or situations involving discrepancies between expectation (adaptation level) and perception are sources of primary, unlearned affect, either positive or negative in
nature" (McClelland et al., 1976, p. 28). Moderate discrepancies lead to a positive reaction, while higher deviation levels cause negative reactions (McClelland
et al., 1976). Since an ambient scent is only one aspect among many others
present in a service or retail environment (Bitner, 1992), its introduction should
lead to a small discrepancy from the adaptation level.

It must be noted that "every sensory event might be considered, at least in
some marginal sense, a discrepancy from some 'expectation' and should therefore lead to some kind of (…) arousal" (McClelland et al., 1976, p. 48). Since an
olfactory signal can be processed fairly straightforwardly to the central nervous
human arousal system without prior cognitive interpretation (Pfaff, 2006), it can
affect individuals even without awareness (Binder et al., 2009; Doty, 2001).
Therefore, also an unconsciously perceived ambient scent will lead to a reaction
on the part of the exposed individual. Importantly, the arousal system determines not only our emotional responses to sensory stimuli, but is also funda-
mental to all cognitive and behavioral reactions (Pfaff, 2006).

Based on our theoretical considerations, we expect that affects, cognitions, and
behavioral intentions of customers will enhance in the short term after the introduction of a pleasurable ambient scent into a servicescape.

H_1: The introduction of a pleasant ambient scent has a positive short-term effect
on (a) affects, (b) cognitions, and (c) behavioral intentions.

2.2.2 Long-term Effects

However, to date, no studies have explicitly addressed the long-term effects of
ambient scents in retail or service environments. We seek to broaden the predominant static view by introducing a long-term perspective of scent influence.
As already suggested by Russell and Lanius (1984), research on environments
should account for a dynamic alteration of individuals' responses to environmental stimuli. We therefore expect that the influence of ambient scents on customers will change over time.

Following the argumentation of optimal arousal theory, expectations of specific stimulation patterns are built on previous experiences, and are subject to change upon new relevant experience. Hence, the adaptation level will adjust over time towards the new stimulation pattern, which will then include the expectation of a pleasurable ambient scent. Therefore, the amount of deviation between the expectation and the de facto perception will level out over time: "[R]eactions (…) appear maximally the first time the discrepancy occurs and with less intensity thereafter because the new experience automatically interacts with the AL, changes it, and thereby reduces the discrepancy" (McClelland et al., 1976, p. 60). This effect is called *habituation* and refers to the phenomenon that a repeated presentation of the same stimulus results in a declining reaction towards it (Mehrabian, 1995; Pfaff, 2006). So, a novel scent stimulus should lead to a stronger response than a repeated ambient scent exposure (Berlyne, 1971; Pfaff, 2006).

This assumption is supported by a recent study by Brüggen et al. (2011), who investigated the short-term and long-term effects of a remodeled servicescape (lightning, colors, layout, etc.) on customers. The results show positive short-term effects on cognitive and behavioral constructs that wear off over time.[20] A similar effect could be expected of a repeated ambient scent exposure, and thus its long-term influence. Consequently, we assume a leveling of the positive short-term effect of a scented servicescape on customers in the long term.

H_2: *The short-term impact of a pleasant ambient scent on (a) affects, (b) cognitions, and (c) behavioral intentions will level out over time.*

2.2.3 Aftereffects

After discussing potential short-term and long-term effects, we also need to understand what happens if an ambient scent is no longer used after some time.

Our argumentation again follows optimal arousal theory, which holds that removing a stimulus represents a change in the stimulation pattern and in turn causes a discrepancy from the adaptation level (McClelland et al., 1976). Following prospect theory, we assume that a negative deviation from a given refer-

---

[20] The authors could neither identify any short-term nor long-term effects on affective measures (Brüggen et al., 2011).

ence point (in this case, a loss of an additional environmental information cue) outweighs a positive deviation (meaning an information gain) (Kahneman & Tversky, 1979). Hence, the loss of an additional environmental information cue – the ambient scent – represents a distinct deviation and therefore causes a negative reaction. As a result, we hypothesize:

H_3: The removal of a longitudinally diffused pleasant ambient scent, will cause a decrease in (a) affects, (b) cognitions, and (c) behavioral intentions.

Figure 4 provides an overview of the three hypotheses. It must be noted that, independent of the existence of any direct effect of olfactory stimulation, reactance might occur within individuals exposed to a scent: If subjects perceive ambient scent use as illegitimate and purposeful manipulation by a service provider, they might generally react negatively (Bradford & Desrochers, 2009; Lunardo, 2012).[21]

Figure 4: Overview of Hypotheses[22]

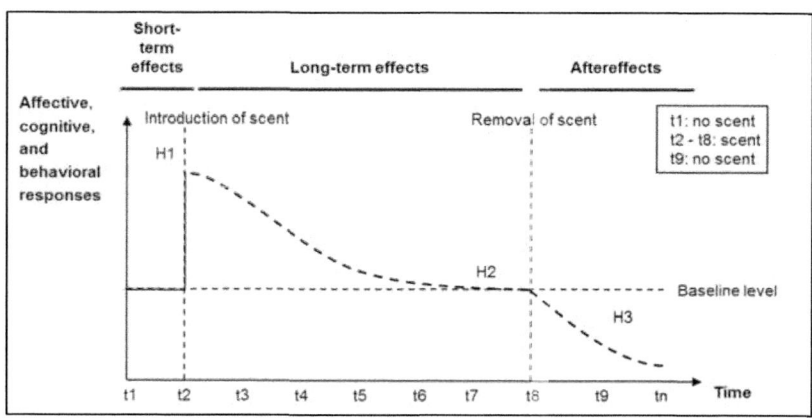

3   Experimental Research Design

To test our hypotheses, we conducted a controlled field experiment over four months from December 2011 to March 2012 with a customer panel of a German railway transportation service company (*Deutsche Bahn*). The aim was to empirically analyze the temporal structure of ambient scent effects; concretely,

---

[21] Remarks concerning reactance see Footnote 16.
[22] Own illustration.

how a systematic diffusion of ambient scents affects customers, who are repeatedly exposed in the long term (commuters), and what aftereffects are observed once upon scent removal. Therefore, we manipulated the olfactory setting via systematic stimulus diffusion in the air conditioning system of all trains in operation on a specific track section in Southern Germany (see Appendix C.1). Overall, 11 trains with at least two coaches each were equipped with scent cartridges that continuously and constantly diffused the experimental stimulus in the servicescape.

With our specific research setting we meet the claim of Capelli and colleagues (2013) that "[h]uman panels (…) are necessary for direct assessment of odour in the field" (p. 731): We applied a pretest/posttest control group design, meaning wave 1 without ambient scent diffusion as baseline measurement, then wave 2 until 8 with ambient scent, and the final wave 9 again without scent use, in order to control for disturbance factors and to allow for reliable measurement of dependent variables. We thus gathered multi-period survey data over 9 waves and compared the development of the targeted affective, cognitive, and behavioral constructs in a customer panel over time using a within-subjects design (Keselman & Algina, 1996; Shadish et al., 2002; Tanguma, 1999). Since we measured the dependent variables at nine points over a relatively long period (4 months), we were able to obtain longitudinal data. Since we additionally varied our independent variable – the scent vs. no-scent condition – we gathered repeated-measures data. By combining the two, our study provided longitudinal repeated-measures data of odor influence with a service customer panel in the field (Capelli et al., 2013; West et al., 2007).

One advantage of the chosen within-subjects approach is that it requires fewer respondents in order to reach high statistical power compared to a between-subjects design, as they serve as their own control. This factor is especially relevant when participants are hard to find, as in our case (see section 3.3) (Girden, 1992; Greenwald, 1976; Keselman & Algina, 1996; Tanguma, 1999). Furthermore, within-subjects designs exclude one main source of error variance attributed to individual variation, which leads to greater statistical power. Removing this variance increases the likelihood that differences over waves are owing to the treatment rather than the participants themselves (Keselman &

Algina, 1996; Tanguma, 1999). The main disadvantage is a potential *practice effect*, either positive or negative in nature, owing to repeated measurements. To minimize a negative influence we offered incentives to increase the motivation to participate and to reduce boredom; to reduce fatigue we set the interval between surveys to one week (Tanguma, 1999). To control for a positive practice effect – so-called *panel conditioning* – we introduced a separate control panel (see separate discussion in section 3.4) (Warren & Halpern-Manners, 2012).

Regarding our hypotheses, short-term effects are defined as changes in dependent variables directly after the introduction of the ambient scent between wave 1 and wave 2, while long-term effects are referred to as changes observed after repeated exposure (waves 3 to 8). Finally, any observed changes after the removal of the olfactory stimulus at the end of wave 8 are considered as aftereffects (wave 9).

Before reporting the results of our experiment, we present the preparation of the experimental scent stimulus, the operationalization of considered dependent variables, the sample composition, and the control for panel conditioning. We conducted all analyses using the statistic software IBM SPSS Statistics 21.

3.1 Experimental Scent Stimulus and Pretest

The scent used in our experiment was purpose-developed by a professional perfumer for the specific servicescape of the railway company based on the management aim of creating an enjoyable train ride experience.[23] The scent composition consists of soothing essences of jasmine, melon, violet leaves, and rosewood, to foster a sense of inner peace and relaxation. Thus, the fragrance was intended to promote a relaxing atmosphere within the train compartments and an enjoyable service experience.

Several pretests were conducted to determine an optimal intensity level of the olfactory stimulus for the main survey (for the questionnaire, see Appendix C.2). Overall, we surveyed 198 different customers at three points in time ($N_{t1} = 86$; $N_{t2} = 50$; $N_{t3} = 62$) with three different associated intensities of the ambient scent (4 scent cartridges per coach; 6 cartridges per coach; 8 cartridges per coach),

---

[23] For a detailed description of scent selection, see M. Girard (2015).

and a maximum of 8 cartridges per train compartment possible owing to technical restrictions of the air conditioning system. Our sample contained more female than male participants (112 vs. 80; 6 missing values), who were between 15 and 64 years old (main age category 15 to 24 years, at 47.4%).[24]

Using the lowest intensity level (4 cartridges), only 25 out of 86 participants (29%) consciously perceived the olfactory stimulus (11 unsupported, 14 supported). When using 6 cartridges, 15 out of 50 (30%) smelled the scent (8 unsupported, 7 supported), and even with the highest intensity level of 8 scent cartridges per train compartment only 13 out of 62 (21%) were able to perceive the scent consciously (5 unsupported, 8 supported).[25] Furthermore, the rating of perceived intensity on a bipolar scale (*very weak/very strong*) (Spangenberg et al., 1996) did not differ significantly between the three intensity levels ($M_{t1} = -.26$, $\sigma = 1.54$; $M_{t2} = .73$, $\sigma = 1.53$; $M_{t3} = .25$, $\sigma = 2.01$; ANOVA: Shapiro-Wilk tests for every intensity level = not significant (n.s.), Levene's test = n.s., $R^2_{adj.} = .024$, $F(2,49) = 1.601$, $p = .213$, $\eta^2 = .064$).

Overall, in none of our pretest scenarios did the intensity reach the so-called collective perception threshold, defined as a conscious, but unspecific scent perception (Binder et al., 2009), where more than 50% of subjects perceive a scent, indicating a collective conscious scent perception (Doty, 1991c; Mücke & Lemmen, 2010).[26] Thus, we have to assume that most of our participants perceived the scent unconsciously. Independent of the consciousness level, an olfactory stimulus can still cause internal and behavioral responses in exposed individuals (Lee & Schwarz, 2012; Levy et al., 1997; Li et al., 2007; Lorig et al., 1991) if the so-called detection (or absolute) threshold has been exceeded. This refers to the lowest intensity when an olfactory stimulation causes an (unconscious) reaction in the autonomous nervous system (Binder et al., 2009). Since at least 21% of our respondents had a conscious smell impression, we assume that our chosen intensity was above the necessary level for absolute detection.

---

[24] Participants were recruited directly on the train and received lipsticks and bags as tokens of appreciation. The first two pretests took place in December 2011 and the third in February 2012.
[25] We found no significant relationship between scent detection and objectively present intensity: $\chi^2(2) = 1.563$, $p > .10$, Cramer's V = .089.
[26] Doty (1991c) suggests a value of 75%, originating from laboratory experiments where test persons had to answer. In his studies, 50% equals scent detection by chance and 100% equals perfect allocation. For field experiments a collective threshold of 50% – as suggested by Mücke and Lemmen (2010) – might be more realistic.

Hence, our scent stimulus will be applied in our main study below the perception but above the detection threshold, using 8 scent cartridges per train compartment.

Furthermore, we used the pretests to ensure that the scent is perceived as appropriate to support an ideal train ride experience (not incongruent), as pleasant (hedonic evaluation), and not as arousing (relaxing). In total, 53 respondents perceived the scent consciously and were therefore able to directly evaluate the olfactory stimulus.

We tested the scent's congruence with the managerial goal of creating an ideal railway experience by asking participants about the scent's appropriateness, as suggested by Bone and Jantrania (1992). Answers were given on a seven-point Likert scale (from *fully disagree* to *fully agree*) to the question *The scent fits my ideal train ride experience*. The results revealed that the scent is rated as not incongruent with an ideal train ride experience (M = 4.04, σ = 1.85), without significantly deviating from its neutral scale midpoint of 4 in an one-sample t-test (t(47) = .156, p = .876, r = 0.023).

The scent's properties were measured with items from Fisher's (1974) environmental quality scale, as suggested and adapted by Bosmans (2006). The scent's arousal level (*relaxed/tense, boring/stimulating, unlively/lively, dull/bright*, Cronbach's α = .733) was perceived as neutral (M = 0.22, σ = 0.96) on a bipolar scale (-3/+3), and did not significantly differ from the scale midpoint of zero (t(38) = 1.418, p = .164, r = .224). Furthermore, the scent was evaluated as pleasant (*good/bad, pleasurable/unpleasurable, comfortable/uncomfortable, good/bad, attractive/unattractive*; α = .951) on a bipolar scale (-3/+3) (M = 0.75, σ = 1.42), with a significant deviation from the scale midpoint of zero (t(39) = 3.348, p = .002, r = .472).

We generally consider a scent's pleasantness as the most important factor, when selecting a scent, even compared to its congruence, since individuals who receive repeated exposure will learn to associate the scent with their de facto train ride experience over time (Engen, 1972).

Finally, we controlled for a low familiarity impression of the stimulus (M = 3.53, σ = 1.53) to avoid any impact of previous scent experiences using the scale of Morrin and Ratneshwar (2003). The respondents rated their familiarity as rather low (*The scent is very familiar to me*) and as deviating significantly from its neutral midpoint of 4 (t(44) = -2.044, p = .047, r = .294) on a seven-point Likert scale (*fully disagree/fully agree*).

Overall, we considered the scent stimulus as appropriate for our experiment.

Owing to the fact that olfactory stimuli can affect exposed individuals independent of their conscious perception, a classical manipulation check seems difficult. We therefore introduced the variable *perceived air quality* (PAQ), as suggested by Girard et al. (2015). PAQ describes an individual's subjective perception of the indoor air quality (von Kempski, 2002), for which an evaluation in a train compartment should be possible for every customer. In contrast, for an evaluation of a specific ambient scent, the same must be perceived consciously. To perform this manipulation check, we included a bipolar single-item measurement of PAQ in the questionnaire: *The perceived air quality in the train is very bad/very good.* As suggested by Kempski (2002) and demonstrated by M. Girard (2015), the introduction of a pleasant ambient scent into a servicescape should improve the perceived air quality in a service environment: After comparing the means of perceived air quality, depending on the three intensity levels (M_{t1} = .72, σ = 1.44; M_{t2} = .84, σ = 1.38; M_{t3} = 1.45, σ = 1.30), we can confirm an improvement of the PAQ evaluation, which is significant between intensity 1 and 3 (p_{Bonf} = .006), as well as 2 and 3 (p_{Bonf} = .063) (ANOVA: Levene's test = n.s., $R^2_{adj.}$ = .042; F(2, 194) = 5.302; p = .006; η^2 = .052),[27] which indicates a successful manipulation of our olfactory research setting with a medium effect size (Cohen, 1988).

3.2 Construct Measurement and Quality Criteria

We chose the affective, cognitive, and behavioral dependent variables included in our research and their corresponding measurement scales, first, based on current servicescape and scent research, as suggested by Brüggen et al. (2011) and, second, based on the results of a previous short-term scent study in cooperation with the same public transportation company (M. Girard, 2015; Girard et al., 2013).

---

[27] As post hoc procedure, we chose the Bonferroni (Bonf) correction, a fairly conservative method that needs equivalence of variance, assed via Levene's test. Normality for the ANOVA is assumed owing to the central limit theorem (Field, 2013).

Though not exhaustive, our questionnaire included three mainly affective constructs (with some cognitive aspects): *service experience* (scale: Brady & Cronin Jr., 2001), *satisfaction* (scale: Specht et al., 2007), and *brand attitude* (scale: Brexendorf et al., 2010).[28] We define service experience as any interaction of a customer with the service company, and consisting of all consciously or unconsciously, positive, neutral, or negative, subjective impressions connected to a service (Berry et al., 2002; Berry et al., 2006; Brady & Cronin Jr., 2001; Brakus et al., 2009; Edvardsson et al., 2010). Satisfaction is defined as "overall evaluation based on the total purchase and consumption experience with a good or service over time" (Anderson et al., 1994, p. 54), and brand attitude refers to a "customers' positive or negative disposition towards a brand, resulting from the overall perception and satisfaction with brand stimuli" (Brexendorf et al., 2010, p. 1150).

We also measured two mainly cognitive constructs, namely perceived *service quality* (scale: Dabholkar et al., 2000), defined as "subjective evaluations of a service experience compared to their expectations for the service" (Berry et al., 2006, p. 48), and perceived *service value* (scale: Harris & Goode, 2004), relating to "the consumer's overall assessment of the utility of a product [or service] based on perceptions of what is received and what is given" (Zeithaml, 1988, p. 14).

Finally, we measured the respondents' *behavioral intentions* (scale: Fichtel, 2009; M. Girard, 2015), which include patronage behavior (repeated and additional service usage), as well as favorable word-of-mouth (Dick & Basu, 1994; Meyer & Oevermann, 1995).

To operationalize and measure each construct, we adjusted existing proven multiple-item scales (see Appendix C.4), which were all to be rated on seven-point Likert scales.[29] Furthermore, following recommendations from previous scent research studies, we considered potential moderators of scent perception, such as gender, age, smoking habits (Chebat et al., 2009; Lehrner et al., 2000;

---

[28] We refer to these as primarily affective constructs (with cognitive components), based on the definitions provided by the authors, who developed the corresponding measurement scales.
[29] The author translated all scales into German, and another researcher translated them back to English. To ensure consistency, conflicts between the original and the retranslated scales were solved by adopting the translated items.

Spangenberg et al., 1996), and also incorporated categorical control variables for education, occupation, direction of commute, and questionnaire receipt.[30]

The questionnaire was successfully pretested with 29 train customers on a different, but similar track section in order to avoid surveying potential panel participants on the target track beforehand.[31] The final questionnaire can be found in Appendix C.3.

It is noteworthy that the ambient scent's diffusion was never made salient during the study in order to not bias the results (Gilbert et al., 1997).[32] After completion of the final wave, all participants received an explanation of the study's intention and the role of the scent stimulus via mail. Nonetheless, we controlled for awareness of the study's background in wave 1 with a direct question asking participants to describe their expectation regarding the intention of the experiment.[33]

Regarding the quality criteria, we evaluated the convergent validity of our scales with separate principal component analysis (PCA) with Varimax rotation, in order to prove one-dimensionality for every construct in each wave. We also checked for reliability and internal consistency of measurements (see Appendix C.4). We had to drop one item of the original behavioral intention scale; all other scales remained unchanged.[34] For reasons of practicality, we will report mean values of quality criteria over all waves per construct going forward (Appendix C.4 provides a detailed overview of all quality criteria). For the PCA, the Bartlett test indicated significant correlations for all constructs, the Kaiser-Meyer-Olkin

---

[30] The participants could choose between questionnaire receipt via mail, email, or personal distribution within the train (see section 3.3).

[31] The pretest took place in early December 2011. The participants received lipsticks and bags as tokens of appreciation. The results revealed that besides behavioral intentions, all constructs were one-dimensional and explained at least 66% of the total variance; factor loadings were all greater than .73, which indicated a high convergent validity of the constructs. Furthermore, all scales had satisfactory reliability (Cronbach's $\alpha \geq .74$). However, the item-to-total correlation of the behavioral intention scale was slightly below the cut-off value (min .49), which militates for an item reduction in the main study.

[32] However, to account for potential feedback or participant questions, we provided an email address and a phone number for both contact persons at the university as well as the service centre of the German railway company. During the whole investigation, we received no requests for information.

[33] We found three cases of awareness of the true study intention (see section 3.3).

[34] As noted in our pretest (see footnote 31), the behavioral intention item *I will very likely contact XYZ again* did not work at all. This might be due to the fact, that in German, *to contact somebody* is strongly associated to personal communication, which is rather unusual in the context of public transportation services.

(KMO) measure verified the sampling adequacy for the analyses with at least KMO = .70, and all measures of sampling adequacy (MSA) values for individual items were greater or equal than .65. All constructs showed only one factor with an eigenvalue greater than one, which explained at least 76% of the variance and factor loadings greater or equal to .74. Both Cronbach's α (≥ .87) as well as the item-to-total correlation (≥ .60) militated in favor of high reliability and internal consistency of our measurements (Field, 2013; Homburg & Giering, 1996). Overall, we considered the quality of our measurements as satisfactory.

For all further analyses, we calculated construct scores as the average value of corresponding items.

3.3 Sample Composition and Descriptive Statistics

The study participants consisted of actual commuters on the selected track section between *Landsberg am Lech* and *Augsburg* in Southern Germany with a daily commute (workdays) of at least 15 minutes one-way in order to allow for enough time to properly fill in the questionnaire.[35] We recruited our respondents either directly in trains of the specific track section or contacted subscribed regular customers via mail in early December 2011. They were informed that their participation would help to enhance the public transportation company's service, and that if they participated in our survey at nine points of time in weekly intervals (public and school holidays excluded), they would receive ticket vouchers and a hiking book worth around €150 as tokens of appreciation.

To allow for enough flexibility, we offered each participant the possibility to choose to either receive the questionnaire via mail, email, or personally provided by student assistants within all trains in operation every Wednesday between 06:00 and 10:00 on the specific track section. Independent of how the questionnaire was distributed, all participants had to complete the survey in the train during their Wednesday morning commute.[36]

---

[35] The *Kneipp®-Lechfeld Bahn* has 10 stops. The distance between *Augsburg* and *Landsberg am Lech* is 40 km, and the overall travel time is about one hour.

[36] For all personally distributed questionnaires, we could monitor that the questionnaire was completed during the train ride. For online participants, we were able to compare the time the survey was completed with the assigned time of commute on the questionnaire. For the mail participants, we could not ensure survey completion on the train. However, as we found no significant difference between the distribution channels (see footnote 39), we assume that mail participants also filled in their questionnaire during their commute.

We contacted 196 interested people, of which N = 105 actually participated in our baseline measurement (wave 1) in mid-December 2011. We had to exclude five respondents from our sample because three subjects knew the study's background, and two subjects answered the questionnaire monotonically. Thus, we started with 100 respondents in the baseline measurement, which over time decreased to 35 subjects, who fully participated in all 9 waves (315 data points).[37] All further analyses refer to these N = 35 panelists. This development represents an average panel mortality of 12% per wave, which is comparably low for a customer panel (Schnell et al., 2005).

Of the 35 participants, 49% were women, 86% non-smokers, and ages ranged from 16 to 61 years, with a mean age of 34.6 (median = 30). The sample can be characterized as highly educated (57% higher degree, 11% A-level, 26% secondary school, and 6% below), in a permanent work relationship (63% employee, 14% civil servant, 3% self-employed, 20% students, pupils, or apprentice), and rural residence (63% commute into the city).[38] The distribution between the three ways of receiving the questionnaire was almost equal (13 train, 11 mail, and 11 email).[39]

Except for one train (18 minutes' delay, which concerns 4 data points), no train used by our commuters on any of the survey days was delayed by more than 10 minutes at its final stop.

---

[37] The base data (N_B = 100) and the final data (N_F = 35) do not significantly differ concerning gender ($\chi^2(1)$ = 0.204, p > .10), smoking habits ($\chi^2(1)$ = 0.454, p > .10), direction of commute ($\chi^2(1)$ = 0.089, p > .10), and type of questionnaire receipt ($\chi^2(1)$ = 0.572, p > .10). The final dataset of N = 35 is significantly older (M_B = 29.06, σ = 12.746; M_F = 34.63, σ = 13.142; Levene's test = n.s.; t(133) = -2.207, p = .029, r = .188), has more participants with higher educational degrees ($\chi^2(5)$ = 10.342, p = .066; Fisher's exact test (FET) = 8.928, p = .081, Cramer's V = .280), and contains more employees and less students/pupils/apprentices ($\chi^2(4)$ = 11.030, p = .017; FET = 12.047, p = .007, Cramer's V = .288) than the base data (N = 100). This kind of panel mortality is not unusual, since participants not interested in the topic tend to fade out of the panel, which can be assumed for younger and less educated respondents (Gross Sobol, 1959).

[38] There was no significant main effect of direction of commute (repeated-measures ANOVA = rANOVA), indicating that the ratings between rural and urban living participants were similar for all constructs.

[39] There was no significant main effect of distribution channel (rANOVA), indicating that the ratings between train, mail, and email participants were similar for all constructs.

3.4   Control of Panel Conditioning

Owing to our experimental design, all trains in operation on our experimental track section were scented between waves 2 and 8 in order to ensure that none of our panelists accidentally used a non-scented train for their commute. So, we were unable to control for panel conditioning through repeated answering with our pretest/posttest design. Hence, we introduced a control panel over three waves on another track section in the same region in Germany (see Figure 5 and Appendix C.1) (Warren & Halpern-Manners, 2012). The recruiting process, instructions, and questionnaire for the control panel were the same as for our main panel.he respondents received the questionnaire via mail and received a ticket voucher worth about €30 as token of appreciation if they participated in all three waves (same timing as the experimental panel waves 1 to 3). In total, 12 female and 13 male respondents aged between <15 to 54 years participated in all waves (N = 25).[40]

Figure 5: Overview of Experimental Setting[41]

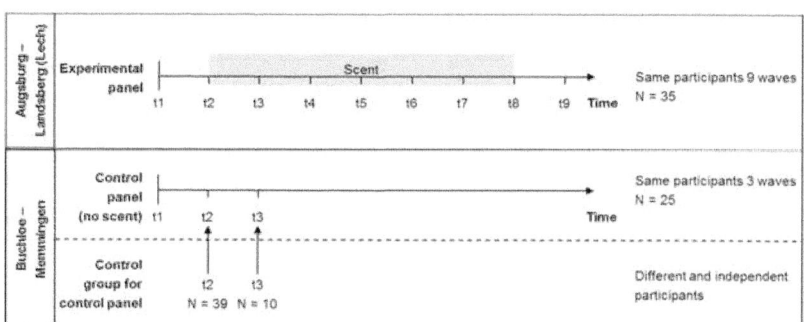

Since Kanji (2006) generally recommends using non-parametric tests if the sample size is smaller than 30, we used Friedman's related-samples ANOVA by ranks (χ^2_F) for the analyses especially, since for some waves the scores of our

---

[40] The control panel (N_C = 25) and the experimental panel (N_E = 35) do not significantly differ concerning gender ($\chi^2(1)$ = 0.002, p > .10), smoking habits ($\chi^2(1)$ = 0.919, p > .10), and direction of commute (to/from the city) ($\chi^2(1)$ = 1.311, p > .10). The control panel is slightly younger than our main panel; 15 to 24 (at 48%) is the main age category ($\chi^2(5)$ = 15.416, p = .004; FET = 14.454, p = .006, Cramer's V = .507). It has significantly less participants with higher educational degrees (60% without A-level, 20% with A-level, and 20% with higher degree) ($\chi^2(4)$ = 11.489, p = .016; FET = 11.191, p = .016, Cramer's V = .438), and is mainly employed as pupil, student, or apprentice (68%), ($\chi^2(3)$ = 15.695, p = .000; FET = 15.536, p = .000, Cramer's V = .511).
[41] Own illustration.

dependent variables satisfaction, service experience, quality, and brand attitude deviate significantly from normal.[42,43] Assuming there is no effect of panel conditioning in the data, the control panel should not differ between waves.

As expected, the dependent variables brand attitude, service value, and behavioral intention did not significantly change over the three waves, and we thus can exclude any practice effect for our main experiment (see Appendix C.5). However, satisfaction ($\chi^2_F(2) = 6.894$, p = .032), service experience ($\chi^2_F(2) = 9.477$, p = .009), and service quality ($\chi^2_F(2) = 7.586$, p = .023) did significantly change over time. Pairwise comparisons were used to follow up on these findings. Results showed a significant positive deviation between waves 1 and 2 in satisfaction (p = .009, r = -.360), as well as service quality (p = .059, r = -.330), and a positive change in service experience between waves 1 and 3 (p = .017, r = -.390). These results indicate a potential panel conditioning with a medium effect (r > |.3|) for the constructs satisfaction, quality, and experience.

We further checked robustness of these results, indicating a potential bias for the main study. We used one time control groups in the same trains as the control panel in waves 2 (N = 39) and 3 (N = 10) (see Figure 5 and Appendix C.1) (Warren & Halpern-Manners, 2012).[44,45] Hence, we were able to compare mean values of the control panel answering the questionnaire for the second or third time, with control group respondents participating for the first time on the same

---

[42] The constructs' quality criteria for the control panel were examined together with the main panel.

[43] We checked for normality using Shapiro-Wilk tests for every construct in each wave; and owing to the fact that normality tests tend to lack power in small samples, we also checked skewness and kurtosis of the distributions reporting their z-scores (Field, 2013).

[44] Participants were recruited directly in trains and received lipsticks and bags as tokens of appreciation. The questionnaire was identical to the one used for the main panel and the control panel.

[45] The control panel ($N_C = 25$) and the control group in wave 2 ($N_{G2} = 39$) do not significantly differ concerning gender ($\chi^2(1) = 0.925$, p > .10), smoking habits ($\chi^2(1) = 0.636$, p > .10), educational degree ($\chi^2(5) = 7.816$, p > .10; FET = 7.650, p > .10), and occupation ($\chi^2(3) = 5.242$, p > .10; FET = 5.095, p > .10). The control group is slightly older than our control panel; 15 to 24 (at 85%) is the main age category ($\chi^2(4) = 10.787$, p = .019; FET = 10.320, p = .017, Cramer's V = .411). In wave 3, the control panel ($N_C = 25$) and the control group ($N_{G3} = 10$) do not significantly differ concerning gender ($\chi^2(1) = 0.945$, p > .10), age ($\chi^2(5) = 8.213$, p > .10; FET = 6.518, p > .10), smoking habits ($\chi^2(1) = 0.875$, p > .10), and educational degree ($\chi^2(4) = 7.029$, p > .10; FET = 6.740, p > .10). The control group contains significantly more retirees (2), and thus differs regarding occupation ($\chi^2(4) = 9.331$, p = .041; FET = 8.158, p = .044, Cramer's V = .516).

track section and at the same points in time using an independent-samples Mann-Whitney U test (see Appendix C.5).[46]

Service quality levels did not differ significantly between both groups (wave 2: U = 416.50, p = .328, r = -.122). The evaluations of service experience also did not deviate significantly between the two samples (wave 3: U = 98.50, p = .339, r = -.164). So, we can assume that the more positive evaluation was not related to a positive practice effect, and thus exclude panel conditioning for our main study. However, in wave 2, our control panel subjects were significantly more satisfied (M = 4.98) than the one-time control group subjects (M = 4.27, U = 353.00, p = .063, r = -.232).

In short, we can exclude panel conditioning for all measured constructs with the exception of satisfaction. As the results indicate only a small effect (r > .1) at a 10% significance level, we decided not to entirely exclude satisfaction from further analyses, but to pay special attention to the results and its interpretation relating to this construct in the main study.

4   Analyses and Results

As longitudinal repeated-measures survey data tend to be serially correlated (also referred to as *autocorrelation* of the residuals, which means that any residuals of two observations are correlated in a sequence of measurements), specific methods of analysis are required (Field, 2013; Nemec, 1996). For each dependent variable, we used a separate repeated-measures ANOVA (rANOVA), which is generally suitable to assess experimental treatments and their influence over time on a decent number of time periods (i.e., number of waves ≤ 10) (Girden, 1992; Nemec, 1996; West et al., 2007).[47] In particular, such procedures are preferable for within-subjects designs with rather small sample sizes in order to analyze effects over time (Voelkle & McKnight, 2012),[48] and has already been used in scent research (Schifferstein et al., 2011).

---

[46] Test is appropriate for small sample sizes and in case of non-normality of the dependent variable (Field, 2013).

[47] For instance, time series analysis are more appropriate to describe the data, investigate trends, develop models, and predict future values for large number of observations (Nemec, 1996).

[48] Even though, considered a fairly traditional and outdated methodology, Voelkle and McKnight (2012) can show in a Monte-Carlo simulation that rANOVAs are more appropriate than latent

Besides normal distribution, which can be assumed owing to the central limit theorem, since our main sample exceeds 30 participants (N = 35) (Field, 2013), independence of observations between participants as well as *sphericity* are prerequisites for rANOVAs (Girden, 1992; Tanguma, 1999). Sphericity means "that variances of differences for all treatment combinations be homogeneous" (Girden, 1992, p. 16), and can be compared to the homogeneity assumption of variance in between-subjects ANOVAs (Field, 2013).[49] Mauchly's test of sphericity assesses this assumption, and in case of violation the Greenhouse-Geisser correction (ε) is reported, which reduces the degrees of freedom (df) (Girden, 1992; Huynh & Feldt, 1976).[50]

The significance level was adjusted to p < .10, as suggested by Bosmans (2006) for the context of scent research. Besides the level of significance, we will also report effect sizes (Fritz et al., 2012; Levine & Hullett, 2002). The rANOVA provides the partial Eta square (ηp^2) in SPSS by default. Additionally, we manually calculated omega square (ω^2), as suggested by Olejnik and Algina (2003), as a robust and appropriate measure for repeated-measures designs, since ηp^2 tends to overestimate effect sizes (Bakeman, 2005). For both estimators, > .0099 equals a small effect, > .0588 a medium, and > .1379 a large effect (Cohen, 1988). A posteriori sensitivity power analysis revealed that with the sample size of N = 35 in our main study, even small effects could be detected (Faul et al., 2007).[51]

---

curve or structural equation modeling approaches when analyzing effects over time in within-subjects designs with small sample sizes.

[49] The rANOVA therefore allows for correlation between repeated data points within one individual and for correlations within one wave between different individuals, as long as the correlations are equal. However, sphericity is not a necessary condition and can be compensated by adjusting the degrees of freedom, in order to make the F-ratio more conservative (Field, 2013; Nemec, 1996).

[50] The Greenhouse-Geisser ε sizes the amount of deviation from homogeneity. "Values of ε range from 1/(k -1) (indicating maximum heterogeneity) to 1 (indicating homogeneity), where k represents the number of levels of the within-subjects factor for a single-factor design" (LaTour & Miniard, 1983, p. 50). In our setting, k equals 9 waves, indicating that ε has to be between .125 < ε < 1. The Greenhouse-Geisser ε is the best estimator if N is more than twice the number of waves (N = 35 > 2*9 waves), and ε is smaller than .75, which is the case for all our analyses (Huynh & Feldt, 1976).

[51] G*Power 3.1.3 analysis with the following input parameters: α = .10, power = .90, group = 1, measurements = 9, correlation among repeated measures = .781 (mean over waves and constructs), non-sphericity correction ε = .606 (mean over constructs). Effects with a effect size of f = .127 could be detected, equaling a small effect size (Cohen, 1988).

4.1   Identified Effects on Affects, Cognitions, and Behavioral Intentions

The results of our experiment are summarized in Table 4.

Table 4: The Effects of Scents on Affective, Cognitive, and Behavioral Intention Measures[52]

			Affective responses			Cognitive responses		Behavioral
			Experience	Satisfaction	Brand	Quality	Value	intention
		N	35	35	34	35	34	33
Mauchly's test of sphericity		χ^2	66.340	88.174	51.545	84.139	71.590	93.630
		df	35	35	35	35	35	35
		p	.001	.000	.037	.000	.000	.000
Greenhouse-Geisser ε			.621	.602	.686	.576	.644	.507
Test of within-subjects effects (rANOVA)		F	7.042	1.071	1.119	2.154	3.878	2.131
		df_M	4.971	4.817	5.489	4.610	5.148	4.054
		df_R	169.022	163.769	181.132	156.738	169.895	129.744
		p	.000	.378	.353	.067	.002	.080
		η_P^2	.172	.031	.033	.060	.105	.062
		ω^2	.149	.004	.002	.032	.057	.018
Tests of within-contrasts[1]	W2-1	F	6.442	1.845	.038	3.131	5.446	.401
		p	.016	.183	.846	.086	.026	.531
		η_P^2	.159	.051	.001	.084	.142	.012
		r	.399	.227	.034	.290	.376	.111
	W3-1	F	17.817	1.124	.004	4.155	12.138	5.200
		p	.000	.297	.948	.049	.001	.029
		η_P^2	.344	.032	.000	.109	.269	.140
		r	.586	.179	.011	.330	.519	.374
	W4-1	F	12.894	.810	1.000	4.327	6.855	3.054
		p	.001	.375	.325	.045	.013	.090
		η_P^2	.275	.023	.029	.113	.172	.087
		r	.524	.153	.171	.113	.415	.295
	W5-1	F	15.580	3.187	1.414	4.336	14.159	4.881
		p	.000	.083	.243	.045	.001	.034
		η_P^2	.314	.086	.041	.113	.300	.132
		r	.561	.293	.203	.336	.548	.364
	W6-1	F	13.826	.213	4.901	3.550	9.035	2.718
		p	.001	.648	.034	.068	.005	.109
		η_P^2	.289	.006	.129	.095	.215	.078
		r	.538	.079	.360	.307	.464	.280
	W7-1	F	19.685	.017	1.089	4.551	10.649	3.651
		p	.000	.897	.304	.040	.003	.065
		η_P^2	.367	.000	.032	.118	.244	.102
		r	.606	.022	.179	.344	.494	.320
	W8-1	F	18.331	.973	.349	6.625	12.827	2.282
		p	.000	.331	.558	.015	.001	.141
		η_P^2	.350	.028	.010	.163	.280	.067
		r	.592	.167	.102	.404	.529	.258
Tests of within-contrasts[2]	W8-9	F	.333			3.381	.088	1.103
		p	.567			.075	.768	.301
		η_P^2	.010			.090	.003	.033
		r	.099			.301	.052	.183

[1] Simple contrast wave 1
[2] Simple contrast wave 9

---

[52] Own illustration.

Mauchly's test indicates that the assumption of sphericity has been violated for all constructs, which is unsurprising, since we manipulated the scent condition and have more than two waves (Girden, 1992).[53] Therefore, degrees of freedom were corrected using Greenhouse-Geisser estimates of sphericity.

The rANOVA results revealed a significant main effect of the ambient scent on the evaluation of service experience $(F(4.971, 169.022) = 7.042$, $p = .000$, $\eta_P^2 = .172$, $\omega^2 = .149$), service quality $(F(4.610, 156.738) = 2.154$, $p = .067$, $\eta_P^2 = .060$, $\omega^2 = .032$), service value $(F(5.148, 169.895) = 3.878$, $p = .002$, $\eta_P^2 = .105$, $\omega^2 = .057$), and behavioral intention $(F(4.054, 129.744) = 2.131$, $p = .080$, $\eta_P^2 = .062$, $\omega^2 = .018$). Overall, we observed a large effect for ambient scent's influence on service experience, and medium to small effects for all other constructs.[54]

To test our hypotheses implying changes of the scent's influence over time compared to the baseline measurement without scent, we conducted follow-up tests with planned comparisons. We used a simple contrast against wave 1 as control category without scent. As suggested by Field (2013), we additionally report the correlation coefficient r besides η_P^2 as more robust effect size measure with $r > .1$ indicating a small, $r > .3$ a medium, and $r > .5$ a large effect size for our pairwise planned contrasts (Cohen, 1988). We first present all results and then discuss them afterwards in the discussion section.

4.1.1 Short-term Effects

As can be seen in Figure 6, which visualizes the development curves of all measured constructs, all dependent variables show at least a marginal increase in wave 2, indicating a positive short-term scent effect (see also Appendix C.6).

---

[53] A scent vs. no-scent condition probably has more variability compared to the two no-scent conditions, which are probably more dependent. Furthermore, "homogeneity among variances of differences is a rare occurrence in studies involving more than two repeated measurements of behavior" (Girden, 1992, p. 18).
[54] For satisfaction and brand attitude, the overall model (rANOVA) is not significant; this is unsurprising, since post hoc tests indicate that there is only one significant wave.

Figure 6: Temporal Structure of Ambient Scents Influence; Deviation from Baseline Measurement[55]

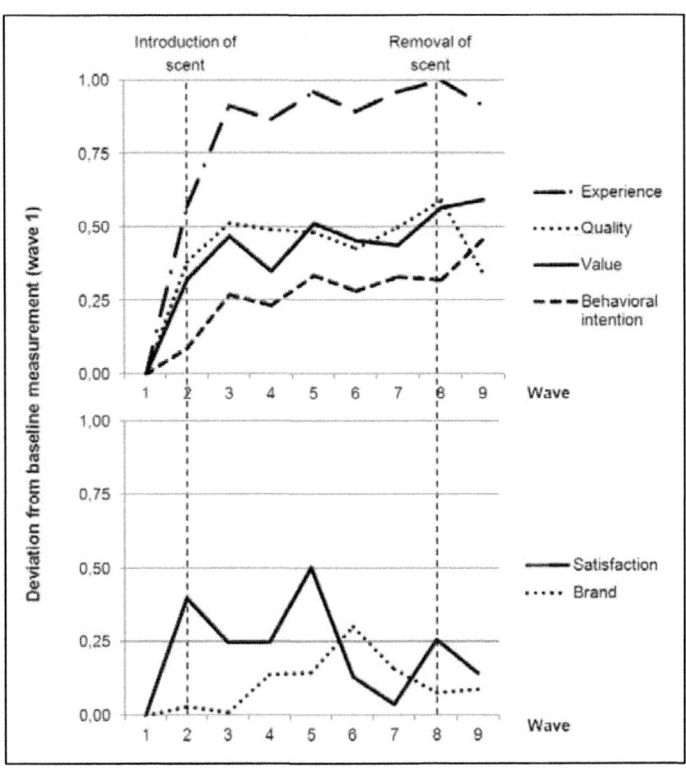

However, as indicated in Table 4, this effect is only significant for service experience $(F(1,34) = 6.442,$ $p = .016,$ $\eta_P^2 = .159,$ $r = .399),$ service quality $(F(1,34) = 3.131,$ $p = .086,$ $\eta_P^2 = .084,$ $r = .290),$ and service value $(F(1,33) = 5.446, p = .026, \eta_P^2 = .142, r = .376).$ The observed effects are medium to large for service experience as well as perceived value, and medium for service quality. These results confirm findings from Girard et al. (2013) in a previous study that investigated ambient scent's impact on railway customers during one-time exposure in a similar experimental setting.

As significance does not imply an effect's importance (Fritz et al., 2012; Levine & Hullett, 2002), it is noteworthy that we could also identify a small effect for

---

[55] Own illustration.

behavioral intention (F(1,32) = .401, p = .531, η_P^2 = .012, r = .111), even though it is not significant.[56] The affective constructs satisfaction and brand attitude show no significantly positive difference between waves 1 and 2 (p > .1), rejecting a short-term impact of ambient scent on these dependent variables. In short, in line with optimal arousal theory, these findings support H_{1b} and provide indication for H_{1c} (observed effect even though not significant). The results emphasize the assumption by Bitner (1992) and Berry et al. (2006) that customers utilize scents as an environmental cue to form their opinion about a service (service quality and value assessment), and then adapt their behavior accordingly. However, concerning affective responses, we found only partial support for H_{1a}, as satisfaction and brand attitude were not directly impacted by the introduction of ambient scents. Only service experience improved significantly after scenting the train compartment for the first time.

4.1.2 Long-term Effects

Overall as can be seen in Figure 6, we identified two development clusters in the long term: First, temporary enhanced evaluations that level out over time, and second, dependent variables that improved under ambient scent's diffusion and stayed at a higher level over the whole observation time.

Belonging to the first cluster, satisfaction and brand attitude show one wave with significant improvement with a small to medium effect that levels out afterwards (see Table 4). Satisfaction reaches its peak in wave 5, with Δ +.504 compared to the baseline (F(1,34) = 3.187, p = .083, η_P^2 = .086, r = .293), and brand attitude is highest in wave 6 with medium effect sizes (Δ +.304, F(1,33) = 4.901, p = .034, η_P^2 = .129, r = .360).

Related to the second cluster, the customers' service experience significantly improved under the influence of scent over time, and remained stable at a high level with a large effect size (.159 < η_P^2 < .367; .399 < r < .606) (see Table 4). Concerning cognitive constructs, we can show an enduring positive effect of scent on service quality and service value. Besides wave 7, perceived service quality differed significantly from the baseline level in a positive way, with a me-

---

[56] Significance can be reached via larger sample sizes. However, the effect size is independent of the number of observations (Fritz et al., 2012).

dium effect size (.084 < η_P^2 < .163; .290 < r < .404) (see Table 4). For perceived service value, we observed positive medium to large effects across all waves (.142 < η_P^2 < .300; .326 < r < .548) (see Table 4). We could further identify a positive long-term effect of scent on behavioral intention, which increased significantly over time (significant effects in waves 3 to 5 and 7) with an overall medium effect size (.087 < η_P^2 < .140; .295 < r < .374) (see Table 4).

In line with optimal arousal theory, we found partial support for H_{2a} as satisfaction and brand attitude leveled out and returned to the baseline prior to the introduction of the scent; however, this was not the case for service experience. Furthermore, we had to reject H_{2b} and H_{2c}, since the repeated scent exposure had an unexpected enduring positive effect on both cognitive and behavioral responses, with medium to large effects. We will elaborate on this finding further in the discussion section.

4.1.3 Aftereffects

To analyze whether there is a significant decrease after the removal of the ambient scent from the train compartments, we compared results between waves 8 and 9 by conducting rANOVAs with planned comparisons using a simple contrast against the last wave as control category without scent.

We found partial support for H_{3b}, since perceived service quality significantly declines from wave 8 to wave 9 (F(1,34) = 3.381, p = .075, η_P^2 = .090, r = .301) (see Table 4). However, we had to reject H_{3a} and H_{3c}, since all other dependent variables showed no significant decline upon scent removal. As Figure 6 revealed, we could at least observe a small reduction for service experience and satisfaction giving indication for H_{3a}. Overall, these results indicate that ambient scent continues to have a positive aftereffect – at least for a while – except for quality assessments.

4.1.4 Moderating Effects

An individual's olfactory acuity (sensitivity to scents), which influences perception of a scent owing to a different perceived scent intensity, is determined by individual moderators such as age, gender, and smoking habits (Doty, 2001; Frye et al., 1990). We assume a moderating effect if there is a significant interaction between the dependent variable and the moderator (Baron & Kenny,

1986; Field, 2013). Therefore, we separately examined potential interaction effects of indicated individual moderators with every construct by integrating them as a between-subjects factor into our rANOVA. This mixed-model design assumes both sphericity and homogeneity of variances (Field, 2013; Girden, 1992).

We found no significant interactions for any dependent variable for age and smoking habits, indicating that ratings and observed effects were similar between ages as well as between smokers and non-smokers for all constructs (see Appendix C.7). However, we found an interaction effect for gender with behavioral intention at a 10% significance level (F(4.276, 132.562) = 2.115, p = .078, η_P^2 = .064), indicating that the long-term impact of ambient scent on behavioral intention differs between male and female participants (see Appendix C.7).[57]

Figure 7: Interaction Graph for Significant Gender x Behavioral Intention Interaction[58]

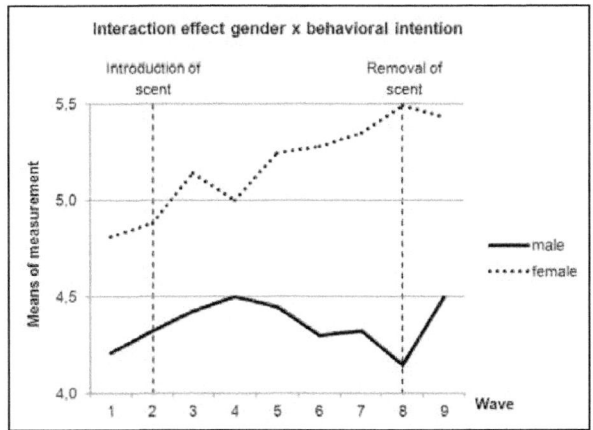

The interaction graph (see Figure 7) reveals that the ambient scent has had an enduring positive effect only on female participants' behavioral intentions, whereas for male respondents, it leveled out over time, as supposed in H$_{2c}$. Be-

---

[57] Levene's test is not significant in any wave; p > .05; Mauchly's test of sphericity: χ^2(35) = 84.842, p = .000, Greenhouse-Geisser ε = .535.
[58] Own illustration.

sides behavioral intention, we found no other significant interactions with gender.[59]

4.2   Additional Qualitative Insights

"Whenever resources allow, field experiments will benefit from including qualitative methods both for the primary benefits they are capable of generating and also for the assistance they provide to the descriptive causal task itself" (Shadish et al., 2002, p. 478). Thus, to gather additional information and insights for the interpretation of our findings (Srnka, 2007), we conducted semi-structured qualitative telephone interviews with 10 panel participants following our main study in August 2012 (Qu & Dumay, 2011).[60,61,62]

We interviewed three male and seven female participants; five of which were employees, two civil servants, two students/apprentices, and one self-employed. The ages ranged from 22 to 62, with an average of 44 years (median = 46.5). The interviews took approximately 10 minutes per person and all interviewees received either a cosmetic product or a cabaret voucher as a token of appreciation.[63]

Our primary focus was to understand whether or not the participants perceived the ambient scent during our study and, if so, how they describe their impressions. We also asked the interviewees about their opinion if the German railway company would permanently introduce ambient scents into its servicescape, and what effects they would expect. Finally, we wanted to know what the company should consider if it was to decide to (permanently) introduce ambient scents in all trains.

---

[59] Also, a combined acuity measure – as suggested by M. Girard (2015) – found null interaction effects (acuity high for females, men between 21 and 50, and non-smokers).

[60] For the sampling, we wrote to all 35 participants in our main study and asked if they declared themselves ready to comply in an additional telephone interview.

[61] Referring to the critical incident approach, we sought to ensure that only prominent facts were remembered. Thus, a timeframe within six months after our study seems appropriate (Keaveney, 1995).

[62] A sample of 10 participants is comparably low for qualitative interviews (Mason, 2010). However, most important for sampling is the concept of *saturation*, for which Guest et al. (2006) could show in an experimentation study that all basic aspects were mentioned after analyzing six semi-structured interviews. After 12 interviews, saturation was reached. For this reason, and as the qualitative study is only intended to gather further information, we consider the sample size of 10 to be appropriate.

[63] For executing and transcribing the interviews (McLellan et al., 2003), the author thanks her student Franziska Ferdinand.

To analyze the interview material, we conducted a qualitative content analysis where two coders classified the responses into a mainly inductive category schema based on the interview guide's structure (Kassarjian, 1977; Krippendorff, 1980; Mayring, 2000).[64] Through a rule-governed approach (Mayring, 2000), with trained and independent judges, we sought to deliver highly objective results (Kolbe & Burnett, 1991). The percentage of agreement (.908) as well as the interjudge reliability (.952) indicate a consistent and reliable allocation of the data by the two judges (Perreault Jr. & Leigh, 1989). As all statements are selective and fully represented by our classification categories and subcategories, we also assume satisfactory content validity (Keaveney, 1995).

Six respondents did not perceive the ambient scent at all during our study, while two participants mentioned scent perception without support, and further two individuals remembered the scent upon request: "I thought, somehow it smells different in here, but then I thought, it's probably just in my imagination."[65] The scent's intensity was evaluated as rather low and not prominent (3 out of 4 participants mentioned the aspect; subsequently abbreviated as 3/4): "When you enter a place, you smell something and then you adapt." The hedonic evaluation of the scent was mixed: Two participants rated it as pleasant, one as neutral, and one as rather unpleasant.

Interestingly, 9 out of 10 respondents were generally in favor of introducing a pleasurable ambient scent into the train compartments if the used substances are non-hazardous to one's health and certified for this purpose. One interviewee concluded that permanent scent diffusion "would really be a great idea." Only one participant opposed ambient scents in general, arguing that he didn't want to be manipulated by scents, but also not by any other marketing instrument, for instance, music.

Overall, all our participants expected exclusively positive effects of ambient scents, even the one who was against a permanent introduction, such as better air quality (6/10), better mood, and more pleasure (5/10), a relaxing effect

---

[64] The category schema contains some deductive categories, which are related to scent characteristics, for instance, quality and hedonic evaluation.
[65] As the interviews were conducted in German, the original quotations were translated by the author to English.

(3/10), and a pleasant atmosphere (2/10). For a permanent scent to be accepted and appreciated by customers, the scent should be pleasant (5/10), fresh (4/10), and have a woody note (2/10), while the intensity should be subtle – as it was in this study (5/10).

During our empirical setting, we deliberately decided not to mention the scent in the questionnaire. During the interview, one customer stated: "But the question [about a scent] was never asked. Hence, I faded it out. (…) The clarification of the study's intention and the scent use afterwards was really a wow factor! (…) I would have never anticipated it. But something was different." Finally, in line with optimal arousal theory, one respondent said: "Once the scent will be integrated permanently, one will get used to it, and the scent becomes commonplace."

5   Discussion and Implications

Based on our quantitative and qualitative results, we now discuss the main findings and then derive implications for practice and academia.

5.1   Main Findings and Theoretical Implications

In this paper, we described a longitudinal field experiment with a German transportation service company, investigating an ambient scent's influence in a servicescape on customers' affects, cognitions, and behavioral intentions. Our findings contribute to marketing and service literature in various ways.

First, longitudinal studies of ambient scents' effects in a marketing context have not yet been addressed in the literature. We introduced a dynamic perspective on short-term, long-term, and aftereffects of ambient scents' influence. Based on the assumptions of optimal arousal theory, we studied olfactory stimulus pattern changes over time.

Second, our investigation confirms the positive customer reactions to pleasurable ambient scents reported in previous studies after a one-time exposure to scent (e.g., Girard et al., 2013; McDonnell, 2007; Morrin & Chebat, 2005) in a real-life transportation service setting.

Third, we found two different patterns of long-term influence of pleasurable ambient scent:

Characterizing the first cluster, our results indicate that the positive effects of scent on the evaluations of satisfaction and brand attitude level out over time, in line with the postulates of optimal arousal theory. As we saw in Figure 6, we observed a jump in the evaluation of satisfaction already in wave 2, which was – however – not significant and might also be due to panel conditioning. Significant effects in the evaluations of satisfaction and brand attitude occurred not directly after the ambient scent's introduction and successively one after another, with brand attitude reaching its peak one wave later than satisfaction. These findings cannot be explained via optimal arousal theory alone, which is why we provide an alternative explanation for the observed time lag: Customer satisfaction is a durable and attitudinal concept that consists of repeated evaluations of a service. The more a customer uses a service, the more stable the evaluation and therefore her/his satisfaction is (Homburg et al., 2006). As we surveyed regular customers, their previously stable satisfaction values might explain why it took several repeated scent exposures to significantly influence their satisfaction assessment. Furthermore, as brand attitude is a result of the brand stimuli's overall perception and satisfaction (Brexendorf et al., 2010; Suh & Yi, 2006), we found a significant change in attitude towards the brand with a time lag following the satisfaction peak. Afterwards, with further scent exposures individuals adapted their expected stimulation pattern and adaptation level accordingly – as postulated by optimal arousal theory.

In the second construct cluster, we found persistent long-term effects, which are not in line with optimal arousal theory: Our assumption is that a positive short-term effect of a scent remains stable if the olfactory stimulus adds 'value' every time the customer is exposed to the ambient scent. In our study, the pleasurable ambient scent had a positive long-term effect on customers' service experience, their perception of service quality, and service value. These results indicate that olfactory cues and thus ambient scents act as an additional sensory experience dimension, as suggested by Brakus et al. (2009). An environmental experience cue "is anything in the service experience the customer perceives by its presence – or absence. If the customer can see, hear, taste, or smell it, it is a clue" (Berry et al., 2006, p. 44). Hence, ambient scents directly contribute to customers' perceived service experience every time they reuse the service (Neslin et al., 2006; Verhoef et al., 2009). By providing additional environmental information, ambient scents present in a servicescape reduce the service's intangibility, and therefore simplify the evaluation of perceived service quality (Goldkuhl & Styvén, 2007; Meyer, 1991; Meyer & Mattmüller, 1987; Zeithaml, 1981). As Zemke and Shoemaker (2007) conclude a "proprietary scent can 'tangibilize' a company's service" (p. 937). Furthermore, environmental experience cues – such as ambient scents – can make a significant contribution to the service value itself, especially for services where the customer stays in a service environment for a while, for instance, in trains, planes, or hotels (Berry et al., 2006).

Overall, we conclude that a scent's long-term influence depends on whether we are looking at a construct that measures a situational evaluation of, for instance, the perceived service experience, the perceived service quality, or the obtained service value. Here, we can observe that scent has a continuous positive influence and contributes to the situational evaluation every time a customer uses a service again. However, if we consider constructs that measure long-term attitudes, such as satisfaction or brand attitude, which evolve over time based on repeated service interactions, we observe that expectations adapt and a scent's influence levels out over time – as predicted by optimal arousal theory.

Fourth, scent's influence on behavioral intention fits neither the first nor the second cluster, because of the significant gender interaction. We found only partial support for our hypothesis that improvements of behavioral intentions under scent influence are only temporary and return to the baseline level over time: This is the case for men, but not for women. As our behavioral intention measure included loyalty and word-of-mouth, we conducted further analyses, which revealed that the gender interaction is predominantly driven by loyalty aspects. This might be due to the facts that, in Germany generally more women than men use public transportation, and that women are more dependent on the service, because they do not own a car, as mostly there is only one car per household, which is used by the male partner. (Bayrisches Landesamt für Statistik und Datenverarbeitung, 2013; Bundesamt für Bauwesen und Raumordnung, 2007). Perhaps, if people do not have a real choice between service providers (so-called *spurious loyalty*) (Dick & Basu, 1994; Javalgi & Moberg, 1997), they rely more on environmental experience cues as additional information source to justify their usage and reduce cognitive dissonance (Bawa & Kansal, 2008; Festinger, 1957).[66] Thus, possibly, women look for more information to reason their choice without actual alternative to reduce cognitive dissonance. Therefore, ambient scents as an environmental cue exert a stronger influence on women's loyalty and behavioral intentions.

Besides the relationship between gender and behavioral intention, we found no evidence for further moderators in scent perception.

Fifth, our findings suggest that a scent's aftereffects persist for a while even after the removal of the stimulus. Except for the customers' service quality assessments, we could not identify any significant drop in the evaluations directly upon scent removal between waves 8 and 9. Our assumption that some aspects of a service are re-evaluated every time a service is used might again explain why, without scent as additional positive environmental cue, the quality evaluation declines. However, the same effect should be expected for service experience and service value. While Figure 6 did show such a decrease for service experience after scent removal, the service value level stays almost flat, and scent seems to have a persisting aftereffect; an effect which we cannot yet explain based on our study.

Finally, our qualitative findings emphasize our empirical findings and show that customers expect mainly positive effects of ambient scents and are mostly in favor of such practices if the stimulus is non-hazardous, subtle, and pleasant.

5.2 Managerial Implications

First, the use of pleasant ambient scents in a servicescape makes sense from a managerial perspective; we found no significant negative consequences of the introduction of a pleasant ambient scent in the short or long term.

The results indicate that a pleasant ambient scent is able to positively influence situational aspects that a customer evaluates every time he or she reuses the service. Thus, service managers can use ambient scents to enhance customer experience, perceived quality, and value – not only temporarily, but continuous-

---

[66] Cognitive dissonance "describes a psychologically uncomfortable state or imbalance that is produced when various cognitions about a thing are not consistent. (…) Cognitive dissonance leads to motivations to reduce dissonance" (Bawa & Kansal, 2008, p. 31).

ly. However, ambient scents are not an adequate marketing instrument to permanently and directly improve attitudes in the long term. A pleasant ambient scent can only temporarily enhance customers' satisfaction and attitudes towards a brand. However, even if a scent might not be able to permanently improve attitudes, a scent can act as a permanent differentiating factor for a brand or a company (Elejalde-Ruiz, 2014), as stated by our respondents: With using a scent "you get an additional sensory link to German Railways (*Deutsche Bahn*)," which might enhance a brand's or a company's recall and recognition value, so that an individual "recognizes the scent when being on the train." Thus, ambient scents can be seen from two perspectives: First, as situational marketing instrument able to directly influence subjective evaluations during a service; second, as brand marketing instrument providing a distinct signal and additional sensory brand dimension. Furthermore, ambient scent might also induce positive behavioral intentions, including loyalty and word-of-mouth, especially in women (or in case of spurious loyalty).

Generally when considering an introduction of ambient scents as a marketing tool, marketers should not only rely on potential short-term commercial benefits of ambient scent diffusion in a one-time test, but also regard the potential influence of long-term scent exposure on their customers in the servicescape: Since some customer reactions to scents weaken over time or might reach a significant level only after a while, companies should carefully test the effects over time before generally introducing ambient scent, since its effects might otherwise be over- or underestimated.

Second, an ambient scent's influence might vary across service industries. Based on the results of our study ambient scents seem especially relevant for services where the situational experience is key, for instance in cinemas or amusement parks, since ambient scents act as an environmental experience cue adding 'value' to the situational experience every time again. As noted, scents might add value to a service only in settings, where a customer remains for a while (e.g., trains, planes, hotels). Thus, we recommend not transferring present findings one-to-one to every other service industry.

Furthermore, qualitative data suggests that customers expect positive effects of ambient scents especially in settings where a negative perceived air quality prevails. A pleasant ambient scent could "compensate for the bad odors" and ensure that the "reek within the train would be neutralized." Therefore, we suggest that the more negative the perceived air quality in a servicescape, the more important pleasurable ambient scents might be.

Third, as the introduction of ambient scents also involves investments and costs (e.g., for developing a scent, installing scent diffusion devices, and maintaining the scent stimulus), managers should control for ambient scents' effects on company performance. While our findings indicate a positive influence on behavioral intentions (at least for female customers), they are not sufficient to prove any impact on monetary economic targets such as sales. Still, our results indicate that, at least directly after the removal of an ambient scent, no negative aftereffects are to be expected, which might jeopardize previous investments. This finding might even open the possibility for ambient scent diffusion as temporary or seasonal marketing tool, to leverage positive ambient scent's effects for a limited period of time, without fearing negative reactions in 'unscented seasons'.

5.3   Limitations and Future Research

As this article constitutes the first study of long-term effects of ambient scents in a marketing context, there are multiple research opportunities.

First, we could identify short-term, long-term, and aftereffects of ambient scents. Still, some issues remain: Not all positive effects occurred immediately after the introduction of an ambient scent. This raises the question of how long it takes to observe an initial influence. Clearly, reaction time differed between constructs: Only service experience, quality, and value showed immediate positive responses – the constructs where ambient scents add value as an environmental cue every time the customer reuses the service and lead to a continuous positive influence. But, why are some effects lagged, like for instance satisfaction, where we did not see an immediate positive effect? We sought to provide an alternative explanation for the observed lag, but still, future research should further investigate this relationship.

Further questions relate to aftereffects: Besides service quality, we found no significant change from wave 8 to wave 9 without scent. So, how enduring are these aftereffects? How long do they persist? With the present study design of only one post-measurement without scent, unfortunately it was not possible to answer these questions. Thus, future research should further investigate the structure and development of scent aftereffects.

Second, the chosen methodology of a field experiment has limitations: Despite the fact that the experiment was cautiously executed and we sought to avoid or control external disturbances, it is impossible to fully exclude any confounding environmental influences (Shadish et al., 2002). In our panel, we controlled for gender, age, smoking habits, education, occupation, direction of commute, and questionnaire reception, which had no significant influence (except for the inter-action effect between gender and behavioral intention). With an average panel mortality of 12% per wave, we were able to retain a high portion of initial re-spondents. However, our base data and our final dataset were not equivalent in age, which is not unusual and should not lead to a systematic bias, as age is not a relevant moderator in our study (Gross Sobol, 1959). We thus do not ex-pect any systematic bias from panel mortality, although we cannot totally ex-clude it.

Furthermore, with our control panel and control group setting, we sought to con-trol for panel conditioning in our main study. Even though the control group in wave 3 was fairly small, we were able to investigate practice effects by choos-ing appropriate analyses. However, as the control surveys took place on anoth-er track section within the same German region, the control customer structure was slightly younger – which should not bias the results, as noted before, but might still have had an influence.

Moreover, no further marketing activities, price increases, or any other media activity associated with the German railway company were conducted during our study. But as our investigation took place in winter, a repetition of the study in summer or over a full year could further control for different weather, temper-atures, and associated olfactory incidents – as suggested by an interviewee: "In winter, when entering the train with wet coats (…) and in summer when people getting in the train – naturally, sweaty." However, field experiment issues such as bad weather as well as train delays might have led to a negative bias of re-sults and an underestimation of scent effects on some waves – thus, ambient scents' effects might be even stronger (or less variable) in other service sectors that are less prone to external influences (e.g., movie theatres).

Third, the field experiment was conducted in one specific service setting. Concerning generalizability of our findings, further studies should investigate whether or not the present results can be replicated and verified also for contexts other than transportation services. Transportation services are mainly characterized by experience qualities (Diederich, 1966), which can be evaluated after utilization. Perhaps the results would be different for services with high search or credence qualities[67] where external environmental information cues become less/more important to evaluate the service (Meyer, 1991; Zeithaml, 1981). We assume that the observed effects get weaker/stronger, the more search/credence qualities a service has (Girard et al., 2015).

Furthermore, with public transportation, we chose a service setting with a fairly negative olfactory situation. Further research should investigate if there are different ambient scent effects depending on the prevailing olfactory situation within the servicescape; whether the air quality is negative, neutral, or positive (Girard et al., 2015). We assume that the observed positive scent influence gets weaker, the better the general olfactory situation is within a servicescape.

Fourth, we investigated scent's effects on selected customer responses, although not exhaustively. It might be interesting to further examine ambient scents' influences on other dependent variables, for instance, emotions or stress. Furthermore, an investigation of *return on marketing decision* might be insightful (Kunz & Hogreve, 2011), to compare the effects of an ambient scent's introduction over time on a company's financial performance. As we chose only one specific track section in Germany, a repetition of the investigation in other German regions or non-German track sections could be interesting to investigate regional or cultural differences.

Finally, in our experimental setting, we only investigated ambient scent's effects on customers. But in service companies, employees are also affected by the servicescape and all present environmental cues (Bitner, 1992, 2000; Girard et

---

[67] Services with a high portion of search qualities are easy to evaluate, as they can be assessed before the service utilization (e.g., retailers); whereas, services with high credence qualities cannot be evaluated after utilization (e.g., professional services such as medical treatments) (Zeithaml, 1981).

al., 2015; Parish et al., 2008). Employees work within a servicescape at least seven hours per day and five days per week (Schweizer et al., 2007). Further research should investigate scents' short- and long-term effects in the workplace, as employees are generally more exposed to ambient scents than customers.

6   Conclusion

The use of ambient scents is an increasing trend in companies. However, managers often utilize ambient scents without really knowing what the long-term impacts of such practices are. Despite the fact that we were able to show only positive or neutral influences of a repeated olfactory exposure on customers, there are several challenges, which should be clarified upfront. Thus, we call for future research to investigate longitudinal emotional, cognitive, and behavioral effects of pleasurable ambient scents in service and retail environments on customers as well as employees in greater depth.

7   Appendix Chapter C

Appendix C.1: Overview of Route Map (Experimental Setting)[68]

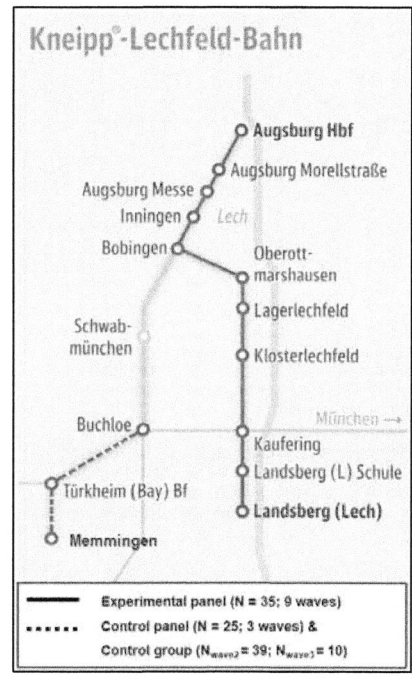

---

[68] Own illustration based on DB Vertrieb GmbH (2014).

Appendix C.2: Pretest Scent Questionnaire (in German)[69]

<table>
<tr><td>**LMU** LUDWIG MAXIMILIANS UNIVERSITÄT MÜNCHEN</td><td>FAKULTÄT FÜR BETRIEBSWIRTSCHAFT MUNICH SCHOOL OF MANAGEMENT</td><td></td><td>MARKETING</td><td>**DB BAHN** Regio Allgäu-Schwaben</td></tr>
</table>

Befragung im Rahmen eines Forschungsprojekts des
Instituts für Marketing der Ludwig-Maximilians-Universität München
in Kooperation mit der DB Regio Allgäu-Schwaben

Nr.	
Zug	
Datum	

Liebe/r Teilnehmer/in,

herzlichen Dank für Ihre Teilnahme an unserer Befragung.

Dieses Forschungsprojekt der Deutschen Bahn Allgäu-Schwaben erfolgt in Kooperation mit dem Institut für Marketing der Ludwig-Maximilians-Universität München und beschäftigt sich mit Ihrer Wahrnehmung der heutigen Reise mit der Deutschen Bahn.

Bitte nehmen Sie sich so viel Zeit wie nötig und beantworten Sie alle Fragen der Reihe nach. Da es um Ihre persönliche Wahrnehmung und Meinung geht, gibt es keine „richtigen" oder „falschen" Antworten. Bitte kreuzen Sie bei jeder Frage diejenige Antwortalternative an, die Ihre Meinung am besten widerspiegelt. Einige Fragen sind sehr ähnlich, jedoch nicht identisch. Beantworten Sie daher bitte alle Fragen vollständig.

Sämtliche Angaben werden selbstverständlich **absolut anonym** behandelt, so dass kein Rückschluss auf Ihre Person möglich ist. Die Daten werden ausschließlich im Rahmen des Forschungsprojekts ausgewertet und dienen wissenschaftlichen Zwecken.

Vielen Dank für Ihre Teilnahme. Mit dem Ausfüllen des Fragebogens helfen Sie uns die Qualität unserer Dienstleistung kontinuierlich zu verbessern.

Für weitere Fragen zu dieser Befragung wenden Sie sich bitte an Fr. Anna Multani, B.Sc., Institut für Marketing, LMU München: multani@bwl.lmu.de / 089 2180 5738

⇨ **Bitte weiterblättern**   1

---

[69] Own illustration.

LMU — LUDWIG MAXIMILIANS UNIVERSITÄT MÜNCHEN — FAKULTÄT FÜR BETRIEBSWIRTSCHAFT MUNICH SCHOOL OF MANAGEMENT — MARKETING — DB BAHN Regio Allgäu-Schwaben

Als erstes würden wir gerne wissen, wie Sie **Ihre heutige Gefühlslage** beschreiben.

Bitte beantworten Sie folgende Fragen nach dem Ausmaß Ihrer Zustimmung zu den einzelnen Aussagen anhand der vorliegenden Gegensatzpaare. Die Ausprägungen an den Rändern stellen hierbei die maximale Zustimmung der entsprechenden Aussagen dar. Bitte kreuzen Sie diejenige Alternative an, die Ihre Meinung am besten widerspiegelt.

1. Wie würden Sie Ihre Gemütslage während der heutigen Bahnfahrt beschreiben?

Sie sind ...

	-3	-2	-1	0	1	2	3	
... unglücklich	O	O	O	O	O	O	O	... glücklich
... genervt	O	O	O	O	O	O	O	... erfreut
... unzufrieden	O	O	O	O	O	O	O	... zufrieden
... betrübt	O	O	O	O	O	O	O	... euphorisch
... gelangweilt	O	O	O	O	O	O	O	... aufgelockert
... nicht froh	O	O	O	O	O	O	O	... froh
... entspannt	O	O	O	O	O	O	O	... aktiv
... gelassen	O	O	O	O	O	O	O	... aufgeregt
... träge	O	O	O	O	O	O	O	... aufgewühlt
... ruhig	O	O	O	O	O	O	O	... unruhig
... schläfrig	O	O	O	O	O	O	O	... munter
... lustlos	O	O	O	O	O	O	O	... angeregt

Nun würden wir gerne von Ihnen wissen, wie Sie die **heutige Luftqualität im Zug** beurteilen.

2. Die Luftqualität im Zug ist heute ...

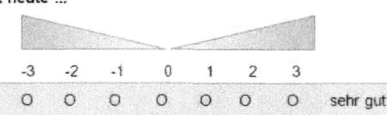

	-3	-2	-1	0	1	2	3	
sehr schlecht	O	O	O	O	O	O	O	sehr gut

3. Haben Sie heute im Fahrgastraum einen besonderen Geruch oder Duft wahrgenommen?

O Ja   O Nein

Falls Sie gerade **NEIN** angekreuzt haben, **fahren Sie bitte mit Frage 4 auf der nächsten Seite** fort.
Falls Sie gerade **JA** angekreuzt haben, **fahren Sie bitte mit Frage 5 auf der nächsten Seite** fort.

⇨ **Bitte weiterblättern**   2

LMU LUDWIG-MAXIMILIANS-UNIVERSITÄT MÜNCHEN | FAKULTÄT FÜR BETRIEBSWIRTSCHAFT MUNICH SCHOOL OF MANAGEMENT | MARKETING | **DB BAHN** Regio Allgäu-Schwaben

4. Nun möchten wir Sie gerne darauf hinweisen, dass dieser Zug heute testweise beduftet wird.

Können Sie diesen Geruch/Duft jetzt im Fahrgastraum wahrnehmen?

    O Ja   O Nein

Falls Sie gerade **JA** angekreuzt haben, fahren Sie bitte mit **Frage 5 fort.**

Falls Sie **NEIN** angekreuzt haben, **überspringen** Sie bitte die nächsten zwei Seiten und fahren mit **Frage 10** auf **Seite 5** fort.

---

5. Wie würden Sie den wahrgenommenen Duft beschreiben?

Bitte beantworten Sie folgende Frage mit eigenen Worten.

_____

_____

Nun würden wir gerne von Ihnen erfahren, wie Sie die **Intensität des Dufts** bewerten.

Bitte beantworten Sie folgende Fragen nach dem Ausmaß Ihrer Zustimmung zu den einzelnen Aussagen anhand der vorliegenden Gegensatzpaare. Die Ausprägungen an den Rändern stellen hierbei die maximale Zustimmung der entsprechenden Aussagen dar. Bitte kreuzen Sie diejenige Alternative an, die Ihre Meinung am besten widerspiegelt.

6. Der Duft im Fahrgastabteil ist ...?

	-3	-2	-1	0	1	2	3	
... sehr schwach	O	O	O	O	O	O	O	... sehr stark

⇨ **Bitte weiterblättern**   3

LMU LUDWIG-MAXIMILIANS UNIVERSITÄT MÜNCHEN | FAKULTÄT FÜR BETRIEBSWIRTSCHAFT MUNICH SCHOOL OF MANAGEMENT

MARKETING **DB BAHN** Regio Allgäu-Schwaben

Nun würden wir gerne von Ihnen erfahren, wie Sie diesen **Duft bewerten**.

Bitte beantworten Sie folgende Fragen nach dem Ausmaß Ihrer Zustimmung zu den einzelnen Aussagen. Zur Bewertung der Aussagen steht Ihnen eine 7-stufige Skala zur Verfügung, auf die Sie Ihre Zustimmung von „stimme überhaupt nicht zu" (linkes Ende) bis „stimme voll und ganz zu" (rechtes Ende) angeben und abstufen können. Bitte kreuzen Sie diejenige Alternative an, die Ihre Meinung am besten widerspiegelt.

7. Bitte bewerten Sie im Folgenden den Duft.

	stimme überhaupt nicht zu		teils teils			stimme voll und ganz zu
	1	2	3	4	5	6	7
Der Duft ist mir bekannt.	O	O	O	O	O	O	O
Ich mag den Duft.	O	O	O	O	O	O	O
Der Duft ist sehr angenehm.	O	O	O	O	O	O	O

8. Welche Eigenschaften hat der wahrgenommene Duft?

Der Duft ist....

	-3	-2	-1	0	1	2	3	
... schlecht	O	O	O	O	O	O	O	... gut
... unangenehm	O	O	O	O	O	O	O	... angenehm
... unbehaglich	O	O	O	O	O	O	O	... behaglich
... negativ	O	O	O	O	O	O	O	... positiv
... nicht ansprechend	O	O	O	O	O	O	O	... ansprechend
... eintönig	O	O	O	O	O	O	O	... vielschichtig
... entspannend	O	O	O	O	O	O	O	... anregend
... langweilig	O	O	O	O	O	O	O	... stimulierend
... leblos	O	O	O	O	O	O	O	... lebhaft

9. Bitte bewerten Sie im Folgenden die Eignung des Duftes.

	stimme überhaupt nicht zu		teils teils			stimme voll und ganz zu
	1	2	3	4	5	6	7
Der Duft passt zu meinem idealen Fahrerlebnis im Zug.	O	O	O	O	O	O	O
Der Duft passt zu meinem Bild der Deutschen Bahn.	O	O	O	O	O	O	O

⇨ **Bitte weiterblättern**   4

Abschließend möchten wir Sie noch um einige Angaben zu Ihrer **Person** bitten.

10. Alter:

 ○ jünger als 15 Jahre ○ zwischen 15 – 24 Jahre ○ zwischen 25 – 34 Jahre

 ○ zwischen 35 – 44 Jahre ○ zwischen 45 – 54 Jahre ○ zwischen 55 – 64 Jahre

 ○ zwischen 65 – 74 Jahre ○ älter als 75 Jahre

11. Geschlecht:

 ○ weiblich ○ männlich

12. Tätigkeit:

 ○ Angestellter ○ Selbstständiger ○ Schüler/Auszubildender/Student

 ○ Rentner ○ Sonstiges: _____

13. Höchster Bildungsabschluss:

 ○ keiner ○ Hauptschulabschluss ○ Mittlere Reife

 ○ Abitur ○ Hochschulabschluss ○ keine Angabe

14. Da wir gerne wissen würden, **wie lange Sie mit dem Zug fahren** würden wir gerne Ihren **Start- und Zielbahnhof erfahren:**

 Startbahnhof: _____ Zielbahnhof:_____

Da bei unserer Studie auch Ihre **Geruchswahrnehmung** relevant ist, würden wir gerne noch Folgendes von Ihnen wissen:

15. Leiden Sie an chronischen Erkrankungen der Atemwege?

 ○ Nein ○ Ja

16. Leiden Sie heute an Erkältung, Schnupfen, Kopfschmerzen, etc.?

 ○ Nein ○ Ja, woran _____

17. Haben Sie Duftempfindlichkeiten bzw. leiden Sie an Allergien auf bestimmte Düfte?

 ○ Nein ○ Ja, welche Düfte: _____

18. Zuletzt würde uns noch interessieren, ob Sie Raucher sind? ○ Nein ○ Ja

---

Sie sind am Ende der Befragung angekommen!
Vielen Dank für Ihre Unterstützung!

5

Appendix C.3: Questionnaire Main Study (in German)[70]

Befragung im Rahmen eines Forschungsprojekts des Instituts für
Marketing der Ludwig-Maximilians-Universität München in
Kooperation mit der DB Regio Allgäu-Schwaben

Fragebogen Nummer 1

Liebe/r Teilnehmer/in,

herzlichen Dank für Ihre Teilnahme an unserer Befragung! Mit dem Ausfüllen Ihres
Fragebogens helfen Sie uns unseren Service kontinuierlich zu verbessern.

- Bitte beantworten Sie alle Fragen der Reihe nach.
- Es gibt keine „richtigen" oder „falschen" Antworten!
- Einige Fragen sind sehr ähnlich, jedoch nicht identisch.
- Sämtliche Angaben werden selbstverständlich **absolut anonym** behandelt.
- Die Daten dienen ausschließlich wissenschaftlichen Zwecken.

**Beantworten Sie daher bitte alle Fragen vollständig, das heißt setzen Sie bitte in
jeder Zeile ein Kreuz!**

Damit wir **Ihren Fragebogen eindeutig zuordnen können**, bitten wir Sie um die Nennung
Ihres **persönlichen Codes**:

Bitte tragen Sie hier Ihren persönlichen Code ein: _____

Der Code setzt sich wie folgt zusammen: Die ersten 2 Buchstaben Ihres Vornamens (Li für Lisa), ersten 2 Buchstaben Ihres Nachnamens
(Sc für Schmidt) und die letzten beiden Ziffern Ihres Geburtsjahres (75 für 1975). Bsp: LiSc75).

Bitte tragen Sie auch das **heutige Datum** ein: _____

In welche **Richtung** fahren Sie gerade?
O Augsburg　　　　　　　　O Landsberg/Lech

Wann ist Ihr Zug abgefahren?
___ : ___ Uhr (hh:mm)

Für weitere Fragen zu dieser Befragung wenden Sie sich bitte an Herrn Abele unter folgender E-Mail-Adresse:
Abele.Maximilian@campus.lmu.de

Bei allgemeinen Rückfragen können Sie sich an den DB Kundendialog wenden unter:
0180 5 99 66 33 - Die Servicenummer der Bahn (14 ct/Min. aus dem Festnetz)

⇨ **Bitte weiterblättern**　　1

---

[70] Own illustration.

DB BAHN
Regio Allgäu-Schwaben

LUDWIG-MAXIMILIANS-UNIVERSITÄT MÜNCHEN — FAKULTÄT FÜR BETRIEBSWIRTSCHAFT MUNICH SCHOOL OF MANAGEMENT

Zu Beginn würden wir gerne wissen, wie **Sie sich im Moment fühlen**.

Bitte beantworten Sie folgende Fragen anhand der vorliegenden Gegensatzpaare. Die Ausprägungen an den Rändern stellen hierbei die maximale Zustimmung zu den entsprechenden Aussagen dar. Bitte kreuzen Sie diejenige Alternative an, die Ihre Meinung am besten widerspiegelt.

1. Wie fühlen Sie sich momentan?

	-3	-2	-1	0	1	2	3	
sehr unwohl	O	O	O	O	O	O	O	sehr wohl

2. Leiden Sie heute an Erkältung, Schnupfen, Kopfschmerzen etc.?

O Nein   O Ja, woran _____

Nun würden wir gerne wissen, wie **Sie die heutige Bahnfahrt beurteilen**.

Bitte beantworten Sie folgende Fragen nach dem Ausmaß Ihrer Zustimmung zu den einzelnen Aussagen. Zur Bewertung der Aussagen steht Ihnen eine 7-stufige Skala zur Verfügung, auf der Sie Ihre Zustimmung von „stimme überhaupt nicht zu" (linkes Ende) bis „stimme voll und ganz zu" (rechtes Ende) angeben und abstufen können. Bitte kreuzen Sie diejenige Alternative an, die Ihre Meinung am besten widerspiegelt.

3. Wie zufrieden sind Sie mit der heutigen Bahnfahrt?

	stimme überhaupt nicht zu		teils teils		stimme voll und ganz zu	
	1	2	3	4	5	6	7
Ich bin insgesamt zufrieden mit der Dienstleistung.	O	O	O	O	O	O	O
Ich bin mit der Leistung der Mitarbeiter zufrieden.	O	O	O	O	O	O	O
Ich bin mit dem Ablauf der Dienstleistung zufrieden.	O	O	O	O	O	O	O

4. Inwieweit stimmen Sie den Aussagen zu Ihrem heutigen Fahrerlebnis zu?

	1	2	3	4	5	6	7
Wenn ich den Zug heute verlasse, habe ich das Gefühl ein schönes Erlebnis gehabt zu haben.	O	O	O	O	O	O	O
Ich glaube, die Bahn versucht mir heute ein angenehmes Erlebnis zu vermitteln.	O	O	O	O	O	O	O
Ich denke, die Bahn weiß, welche Erlebnisse sich ihre Kunden wünschen.	O	O	O	O	O	O	O

5. Wie beurteilen Sie die Qualität der heutigen Bahnfahrt?

	1	2	3	4	5	6	7
Die Qualität der Bahnfahrt ist allgemein sehr gut.	O	O	O	O	O	O	O
Der Bahnfahrt ist qualitativ hochwertig.	O	O	O	O	O	O	O
Die Bahnfahrt hat einen hohen Qualitätsstandard.	O	O	O	O	O	O	O
Die Qualität der Bahnfahrt ist ausgezeichnet.	O	O	O	O	O	O	O

⇨ **Bitte weiterblättern**   2

LMU LUDWIG-MAXIMILIANS-UNIVERSITÄT MÜNCHEN | FAKULTÄT FÜR BETRIEBSWIRTSCHAFT MUNICH SCHOOL OF MANAGEMENT

MARKETING **DB BAHN** Regio Allgäu-Schwaben

Nun würden wir Ihnen gerne einige **allgemeine Fragen zur DB Regio Allgäu-Schwaben** stellen.

6. Wie sehr stimmen Sie den folgenden Aussagen zu?

	stimme überhaupt nicht zu			teils teils			stimme voll und ganz zu
	1	2	3	4	5	6	7
Die Deutsche Bahn ist ansprechend.	O	O	O	O	O	O	O
Die Deutsche Bahn ist anziehend.	O	O	O	O	O	O	O
Ich mag die Deutsche Bahn.	O	O	O	O	O	O	O

Nun würden wir gerne erfahren, wie Sie das **Preis-Leistungsverhältnis** der DB Regio Allgäu-Schwaben einschätzen.

7. Inwieweit entsprechen folgende Aussagen Ihrer Meinung zum Preisleistungsverhältnis der Deutschen Bahn?

	1	2	3	4	5	6	7
Die Angebote der Bahn bieten ein hervorragendes Preisleistungsverhältnis.	O	O	O	O	O	O	O
Die Dienstleistungen der Bahn bieten einen hohen Nutzen.	O	O	O	O	O	O	O
Ich bin zufrieden mit der Leistung, die ich bei der Bahn für mein Geld bekomme.	O	O	O	O	O	O	O
Die Dienstleistungen, die ich bei der Bahn nutze, sind jeden Cent wert.	O	O	O	O	O	O	O

Nun würden wir gerne erfahren, ob Sie die Deutsche Bahn auch außerhalb Ihrer Pendler-Zeit **wieder nutzen und/oder weiterempfehlen** würden.

8. Bitte bewerten Sie, inwieweit folgende Verhaltensabsichten auf Sie zutreffen.

	stimme überhaupt nicht zu			teils teils			stimme voll und ganz zu
Ich denke, ich werde...	1	2	3	4	5	6	7
... nicht zögern den Regionalverkehr der Bahn außerhalb meiner Pendlerzeiten wieder zu nutzen.	O	O	O	O	O	O	O
... anderen positiv von der Dienstleistung der Bahn zu berichten.	O	O	O	O	O	O	O
... die Bahn weiterempfehlen, falls jemand eine Reise plant.	O	O	O	O	O	O	O
... die Bahn außerhalb meiner Pendlerzeiten wieder nutzen.	O	O	O	O	O	O	O
... die Bahn anderen weiterempfehlen.	O	O	O	O	O	O	O
... sehr wahrscheinlich wieder mit der Bahn in Kontakt treten.	O	O	O	O	O	O	O

⇨ **Bitte weiterblättern**   3

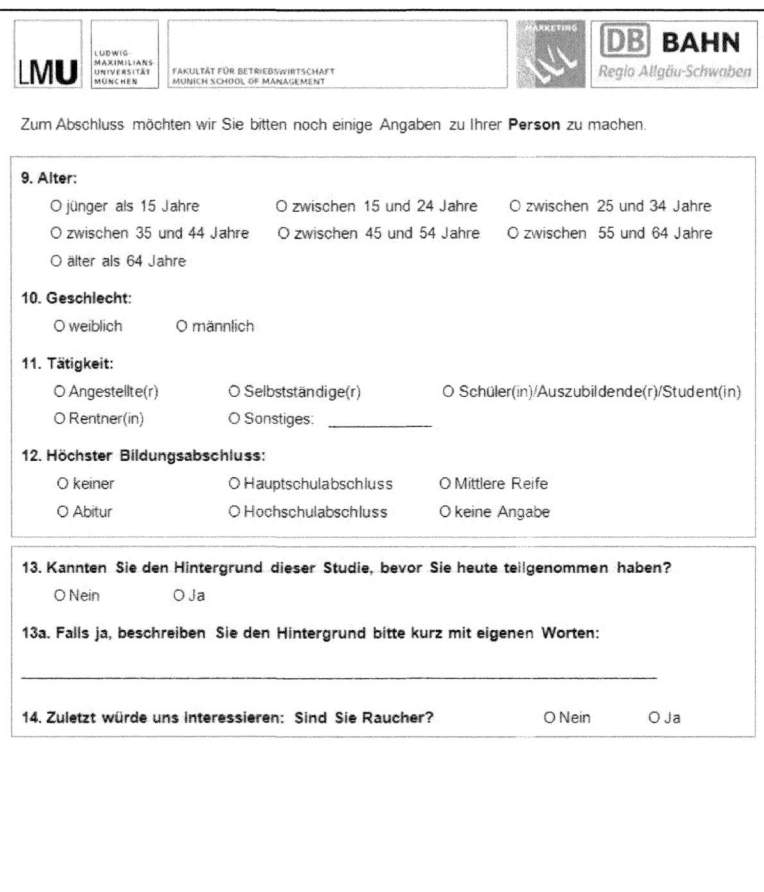

Zum Abschluss möchten wir Sie bitten noch einige Angaben zu Ihrer **Person** zu machen.

9. Alter:

O jünger als 15 Jahre   O zwischen 15 und 24 Jahre   O zwischen 25 und 34 Jahre

O zwischen 35 und 44 Jahre   O zwischen 45 und 54 Jahre   O zwischen 55 und 64 Jahre

O älter als 64 Jahre

10. Geschlecht:

O weiblich   O männlich

11. Tätigkeit:

O Angestellte(r)   O Selbstständige(r)   O Schüler(in)/Auszubildende(r)/Student(in)

O Rentner(in)   O Sonstiges: _____

12. Höchster Bildungsabschluss:

O keiner   O Hauptschulabschluss   O Mittlere Reife

O Abitur   O Hochschulabschluss   O keine Angabe

13. Kannten Sie den Hintergrund dieser Studie, bevor Sie heute teilgenommen haben?

O Nein   O Ja

13a. Falls ja, beschreiben Sie den Hintergrund bitte kurz mit eigenen Worten:

_____

14. Zuletzt würde uns interessieren: Sind Sie Raucher?   O Nein   O Ja

Sie sind am Ende der Befragung angekommen!
Vielen Dank für Ihre Unterstützung!

4

Appendix C.4: Constructs, Measurement Items, and Quality Criteria[71]

Item #	Questionnaire item wording	Factor Loading mean [min-max]	KMO mean [min-max]	Bartlett sig.	MSA mean [min-max]	Variance Explained mean [min-max]	Cronbachs α mean [min-max]	Item to Total mean [min-max]
Affective Responses								
Service Experience (Brady and Cronin 2001)								
1	When I leave the railway, I usually feel that I had a good experience.	.881 [.773-.947]						.741 [.513-.882]
2	I believe German Railways tries to give me a good experience.	.938 [.868-.976]	.702 [.658-.763]	sig.	.648 [.595-.712]	.823 [.670-.926]	.883 [.746-.957]	.946 [.662-.941]
3	I believe German Railways knows the type of experience its customers want.	.898 [.812-.967]						.769 [.567-.923]
Satisfaction (Specht, Fichtel and Meyer 2007)								
1	Overall, I am satisfied with the service of German Railways.	.928 [.900-.959]						.826 [.776-.902]
2	I am satisfied with the employees' performance of German Railways.	.866 [.789-.918]	.699 [.641-.745]	sig.	.649 [.587-.716]	.805 [.734-.847]	.874 [.807-.904]	.710 [.562-.807]
3	I am satisfied with the process of service delivery of German Railways.	.837 [.853-.929]						.760 [.670-.828]
Brand Attitude (Brexendorf et al. 2010)								
1	The German Railway is attractive.	.914 [.842-.983]	.731 [.714-.742]	sig.	.684 [.638-.716]	.835 [.746-.930]	.897 [.929-.962]	.831 [.678-.960]
2	The German Railway is desirable.	.889 [.781-.946]						.790 [.697-.881]
3	The German Railway is likeable.	.889 [.767-.764]						.786 [.681-.917]
Cognitive Responses								
Perceived Service Quality (Dabholkar, Sheperd and Thorpe 2000)								
1	The overall railride quality is excellent.	.912 [.859-.959]						.843 [.751-.923]
2	The railride is of a very high quality	.939 [.919-.964]	.815 [.752-.871]	sig.	.796 [.724-.864]	.861 [.811-.900]	.944 [.920-.961]	.888 [.847-.927]
3	The railride has a high quality standard.	.943 [.915-.983]						.897 [.851-.963]
4	The railride is superior in every way.	.916 [.860-.937]						.851 [.758-.897]
Perceived Service Value (Harris and Goode 2004)								
1	German Railways services are excellent value for the money.	.905 [.855-.946]						.816 [.731-.889]
2	German Railways services are excellent value.	.744 [.583-.849]						.604 [.422-.723]
3	I am happy with the value for money I get at German Railways.	.911 [.839-.952]	.807 [.758-.837]	sig.	.766 [.697-.806]	.758 [.705-.797]	.891 [.856-.915]	.826 [.709-.900]
4	The services I purchase from German Railways are worth every cent.	.907 [.836-.942]						.817 [.703-.877]
Behavioral Intention								
Behavioral Intention (Fichtel 2009, Girard, M. 2015)								
1	I will not hesitate to use German Railways again out of my commute.	.863 [.828-.915]						.798 [.749-.870]
2	I will tell positive things about German Railway in general to other people.	.912 [.852-.960]						.852 [.758-.932]
3	I will recommend German Railway, if somebody is planning a journey.	.920 [.888-.954]	.798 [.752-.853]	sig.	.744 [.679-.835]	.809 [.759-.869]	.938 [.918-.961]	.862 [.807-.917]
4	I will use German Railways again besides my commute.	.876 [.823-.904]						.816 [.734-.856]
5	I will recommend German Railways to other people.	.922 [.878-.958]						.866 [.792-.930]

Note: sig. = significant in each wave with p = .000

---

[71] Own illustration.

Appendix C.5: Tests of Panel Conditioning[72]

	Construct		Experience	Satisfaction	Brand	Quality	Value	Behavioral intention
Related-samples Friedman's ANOVA by ranks	N		25	25	24	25	25	23
	df		2	2	2	2	2	2
	χ^2_F		9.477	6.894	1.026	7.586	2.523	2.810
	p		.009	.032	.599	.023	.283	.245
	W1-W2	χ^2_F	-.600	-.720		-.660		
		Adj. p	.102	.033		.059		
		r	-.300	-.360		-.330		
	W1-W3	χ^2_F	-.780	-.360		-.600		
		Adj. p	.017	.609		.102		
		r	-.390	-.180		-.300		
	W2-W3	χ^2_F	-.180	.360		.060		
		Adj. p	.525	.609		1.000		
		r	-.090	.180		.030		
Independent-samples Mann-Whitney U test	W2 control panel vs. group	U		353.000		416.500		
		p		.063		.328		
		r		-.232		-.122		
	W3 control panel vs. group	U	98.500					
		p	.339					
		r	-.164					

---

[72] Own illustration.

Appendix C.6: Construct Means, Standard Deviation, and Differences per Wave[73]

Wave / Construct		Experience	Satisfaction	Brand	Quality	Value	Behavioral intention
1	Mean	2.838	4.857	4.020	3.800	3.331	4.521
	Standard deviation	1.049	1.253	1.107	1.254	1.015	1.404
2	Mean	3.410	5.256	4.050	4.186	3.654	4.612
	Standard deviation	1.472	1.224	1.280	1.409	1.201	1.381
	Difference	0.572	0.399	0.030	0.386	0.323	0.091
3	Mean	3.752	5.105	4.029	4.314	3.801	4.794
	Standard deviation	1.382	1.212	1.093	1.196	1.224	1.336
	Difference	0.914	0.248	0.009	0.514	0.470	0.273
4	Mean	3.705	5.105	4.157	4.293	3.684	4.758
	Standard deviation	1.333	1.283	1.075	1.181	1.105	1.213
	Difference	0.867	0.248	0.137	0.493	0.353	0.237
5	Mean	3.799	5.361	4.167	4.286	3.846	4.861
	Standard deviation	1.528	1.071	1.201	1.356	1.214	1.448
	Difference	0.961	0.504	0.147	0.486	0.515	0.340
6	Mean	3.734	4.990	4.324	4.229	3.787	4.806
	Standard deviation	1.525	1.298	1.311	1.417	1.282	1.349
	Difference	0.896	0.133	0.304	0.429	0.456	0.285
7	Mean	3.800	4.895	4.177	4.300	3.772	4.855
	Standard deviation	1.564	1.295	1.321	1.373	1.256	1.503
	Difference	0.962	0.038	0.157	0.500	0.441	0.334
8	Mean	3.838	5.114	4.098	4.393	3.897	4.842
	Standard deviation	1.573	1.272	1.306	1.433	1.243	1.646
	Difference	1.000	0.257	0.078	0.593	0.566	0.321
9	Mean	3.753	5.000	4.108	4.143	3.926	4.982
	Standard deviation	1.516	1.318	1.433	1.482	1.366	1.424
	Difference	0.915	0.143	0.088	0.343	0.595	0.461

---

[73] Own illustration.

Appendix C.7: Results of Moderator Analyses[74]

Construct	Mauchly's test of sphericity			Green-house-Geisser	Test of within-subjects effects (rANOVA)				
	χ^2	df	p	ε	F	df_M	df_R	p	η_P^2
Interaction effect age x wave									
Experience	63.226	35	.003	.599	.748	19.18	143.84	.764	.091
Satisfaction	83.939	35	.000	.565	1.466	18.07	135.50	.111	.164
Brand	52.023	35	.034	.678	.914	21.69	157.26	.575	.112
Quality	77.250	35	.000	.549	1.015	17.58	131.83	.447	.119
Value	64.283	35	.002	.642	.904	20.54	148.89	.585	.111
Behavioral intention	82.846	35	.000	.527	1.137	16.88	118.15	.328	.140
Interaction effect smoking x wave									
Experience	66.116	35	.001	.611	.839	4.89	161.31	.522	.025
Satisfaction	95.926	35	.000	.582	.821	4.66	153.68	.529	.024
Brand	49.015	35	.061	-	.931	8.00	256.00	.491	.028
Quality	84.122	35	.000	.571	.412	4.57	150.67	.824	.012
Value	71.888	35	.000	.635	.447	5.08	162.44	.817	.014
Behavioral intention	96.987	35	.000	.490	.784	3.92	121.51	.536	.025
Interaction effect gender x wave									
Experience	67.297	35	.001	.604	1.152	4.83	159.48	.335	.034
Satisfaction	89.050	35	.000	.584	1.015	4.67	154.16	.408	.030
Brand	49.171	35	.059	-	1.477	8.00	256.00	.166	.044
Quality	83.253	35	.000	.568	.785	4.55	150.05	.551	.023
Value	68.998	35	.001	.644	.272	5.15	164.74	.932	.008
Behavioral intention	84.842	35	.000	.535	2.115	4.28	132.56	.078	.064

---

[74] Own illustration.

D. THE IMPACTS OF AMBIENT SCENTS IN THE WORKPLACE: A QUALITATIVE INVESTIGATION

ANNA L. GIRARD

0   Abstract

Employees mostly rate air quality in the workplace as dissatisfying. Research on the indoor environmental quality of workplaces has shown that poor air quality has a considerable negative effect on employees, which is persistent. One possibility to enhance indoor air quality is to diffuse pleasant ambient scents, which have been used for medical purposes for centuries. However, little research has investigated the effects of ambient scent on employees in their workplace. This study provides first insights, applying a diary study with seven employees over five workdays, and sheds light on the questions how employees perceive ambient scents in their workplace and how they are affected. The results show that self-reported effects of ambient scents divert from indirectly measured influences of scent. While the employees predominantly reported that the pleasurable scent had no effect, the scent appeared to unconsciously lead to higher enjoyment, less negative emotions at work, and reduced stress levels. We were able to find mixed influences on employees' perceived capabilities. Furthermore, employees' general attitude towards the presence of ambient scent seems to influence their evaluation of scent in the workplace. Overall, we conclude that olfactory cues present in workspaces do always affect employees. So, even if no ambient scents are actively diffused, poor indoor air quality will lead to negative reactions, which could at least be neutralized by the diffusion of pleasurable ambient scents. Employers should be aware of the importance of perceived air quality to their workforce, and might thus consider air quality as an actively manageable company resource.

Keywords: ambient scent, indoor air quality, work environment, employee behavior, qualitative diary study, field theory

Acknowledgments: The author acknowledges valuable comments by Prof. Dr. Ingo Weller, and thanks Dr. Marc Girard for his general feedback and support as second coder in qualitative content analysis.

1   Introduction

Most employees spend seven to eight hours per day and five days per week in their workplace (Schweizer et al., 2007). Companies should thus seek to provide a pleasant work atmosphere. Research on the overall indoor environmental quality of workplaces has shown that the associated indoor air quality (IAQ) has a considerable impact on employee performance, well-being, and satisfaction (Kim & de Dear, 2012). However, most employees rate the air quality in their workplace as dissatisfying (Frontczak & Wargocki, 2011). Thus, IAQ is especially associated with negative health effects on employees, which are estimated to cause direct medical costs of about US$15 billion annually and a loss of about 150 million workdays per year (Burroughs & Hansen, 2004). One possibility to enhance indoor air quality is the introduction of pleasurable ambient scents (von Kempski, 2002), which have been used for medical purposes for centuries (Herz, 2009). Hence, the diffusion of pleasurable ambient scents might exert positive influences on employees in their workplaces.

Yet, to date, ambient scents are used by more and more companies within their facilities to predominantly create a favorable atmosphere and extraordinary experiences for customers (Goldkuhl & Styvén, 2007; Kroeber-Riel & Weinberg, 2003). Even if used with the intention to influence customers, ambient scents in a business environment usually also affect employees' perceptions (Bitner, 1992; Girard et al., 2015; Parish et al., 2008). Therefore companies might benefit in multiple ways: Besides positive effects on customers, the impact on employees should also be considered, for instance via employee performance or satisfaction enhancements, which might provide an additional economically effective edge in the marketplace (von Kempski, 2002).

Within the past 20 years, the influence of ambient scents has received considerable attention from academics in general. Empirical studies have researched the effects of scents on emotional (Knasko, 1992; Rotton, 1983), cognitive (Baron, 1990; Baron & Bronfen, 1994; Baron & Thomley, 1994), and behavioral reactions by individuals in a variety of work-related contexts (Barker et al., 2003; Ho & Spence, 2005; Raudenbush et al., 2002; Sakamoto et al., 2005). However, research lacks empirical insights into the effects of scents in real-life business environments, since research has to date mainly focused on laboratory experiments. To our best knowledge, there is no empirical evidence of ambient scents' effects on actual employees in their workplace, despite the fact that individual responses to scents are so complex that Kirk-Smith and Booth (1987) conclude: "This almost eliminates the possibility that traditional 'sterilized' types of laboratory-based experimentation will uncover the responses that are normal in the 'real-life' situation" (p. 164).

Thus, there is both an empirical as well as a managerial need for a more in-depth understanding of the effects of scents on employees in their workplace. This paper seeks to contribute to the limited, but emerging body of research by addressing the following central research question: What are the emotional, cognitive, and behavioral consequences of a scented workplace on employees?

To answer this question, we build on Lewin's (1946) field theory as a theoretical basis, explaining why ambient scents will generally affect employees in their work environment. Second, we will review existing studies from environmental quality research as well as scent research and seek to integrate these different perspectives. Most importantly, we will discuss the results of a qualitative diary study, where we accompanied employees over five workdays in order to investigate emotional, cognitive, and behavioral responses to a scented workplace. With our setting over several days, we are able to provide a more realistic view on how scents affect employees in their work environment eight hours per day, five days per week. To our best knowledge, our investigation represents the first attempt to investigate the effects of scents over time. The paper concludes with management implications, further research opportunities, and a final conclusion.

2   Theoretical Foundation: Field Theory

Nowadays, field theory is mostly used in the context of organizational and change-related topics (Burnes & Cooke, 2013), but also in workplace research (Elie-Dit-Cosaque et al., 2012), and service research (Houston et al., 1998).

Kurt Lewin's (1946) field theory posits that behavior (B) is determined by an interaction between a person (P) and his or her situational environment (E), described by the following equation $B = f(P,E)$; meaning that a person's inherent characteristics can affect his perception of the environment, but at the same time an individual's perception depends on the specific environmental situation

(K. Lewin, 1946; Lück, 1996). Hence, the person itself and his environment are considered inseparable and mutually dependent. Lewin (1946) calls the individual's holistic perception of a present situation the *life space* (M. A. Lewin, 1998).[75] In this sense, behavior is a result of psychological forces in an individual's life space; changes in behavior relate to changes in these psychological forces as well as interpretations of the environment. Even if only one element among many in an environment is altered, it will still affect the situation as a whole, and thus the individual's behavior (Burnes & Cooke, 2013; Gold, 1992). Lewin defines behavior not only as actually observable behavior (e.g., talking), but also as individual emotions and cognitions (K. Lewin, 1946). Thus, external cues within the environment – such as ambient scents – might lead to psychological and/or physical reactions (Houston et al., 1998; M. A. Lewin, 1998).

Figuratively, the introduction of ambient scent as external environmental cue in the workplace, defined as "the day-to-day social and physical environment in which you currently do most or all of your work" (Amabile et al., 1996, p. 1165), should affect individual emotions, cognitions, and behavior (f(P,E) → B). Furthermore, a person's scent acuity (i.e., sensitivity to scents), which depends on various individual characteristics, such as gender, age, and/or previous experiences, determines his or her perception of an ambient scent (P → E) (Brand & Millot, 2001; Doty, 1991a; Gulas & Bloch, 1995; Poellinger et al., 2001). The same scent might thus be perceived differently by different persons and consequently trigger different responses (Girard et al., 2015). At the same time, whether or not individuals actually perceive an olfactory stimulus also depends on the specific situation; for instance, air pressure, temperature, or other elements present in an environment (E → P) (Gulas & Bloch, 1995; Knoblich et al., 2003; Krishna, 2012; Petzold, 1983). Therefore, in line with field theory, we conclude that the introduction of an ambient scent in the workplace should alter individual emotions, cognitions, as well as behavior. We derive the following research question:

RQ1. Does the introduction of an ambient scent in the workplace lead to internal and behavioral responses of employees?

---

[75] Individuals have separate life spaces for different occasions, such as work and home (Burnes & Cooke, 2013).

Furthermore, field theory suggests that "the life space includes only those aspects of the environment that are perceived at some level, either consciously or unconsciously, by the individual" (Burnes & Cooke, 2013, p. 412). In other words, external environmental cues that the employees are not necessarily aware of will also affect their life (work)space (M. A. Lewin, 1998; Wheeler, 2008). This aspect is especially relevant for scent perception, since the human brain can process olfactory cues consciously and unconsciously. However, scents are able to affect individuals independently of the level of consciousness (Li et al., 2007; Lorig et al., 1991; Slotnick & Weiler, 2009). Figuratively, we are interested if there is any difference between self-reported influence of conscious scent perception – what do the employees think how an ambient scent affects them – and scent effects they are unaware of (either unconscious or conscious scent perception with an observable, but not reported effect). Therefore, we derive the following research question:

RQ2. Is there a difference between self-reported and observed scent effects on employees?

Field theory also has a dynamic perspective (Burnes & Cooke, 2013), which emphasizes the *principle of contemporaneity*, which means that "any behavior or any other change in the psychological field depends only upon the psychological field at that time" (K. Lewin, 1943, p. 294). In other words, previous and future experiences or events do not influence a life space per se; rather, a previous experience or event changes the present situation, which leads to the individual's reaction. In that sense, scent perception is further characterized by specific time effects called *adaptation* and *habituation*, which are based on previous scent experiences and lead to sensitivity reductions in the present. Adaptation, as a stimulus-induced sensitivity reduction, leads to a decrease in perceived intensity owing to a continuous exposure to a specific scent, for instance during office hours. In contrast, frequent exposure to a specific scent, for instance for several workdays, will induce the effect of habituation. Habituation relates to an experience-based sensitivity reduction that can lower perceived intensity to a completely unconscious perception after repeated exposure to one specific scent (Mücke & Lemmen, 2010; Poellinger et al., 2001). Thus, our third research question is:

RQ 3. Does an employee's perception of an ambient scent vary over time?

As an initial step towards answering our research questions, we will first discuss existing empirical findings.

3 Existing Empirical Findings

The influence of indoor air has so far been investigated from two different perspectives, which will be discussed in the following.

3.1 The Effects of Indoor Air Quality in the Workplace

"The internal physical environment within offices has been given very little attention and is one of the most vaguely understood aspects of management and organizational behavior" (Davis, 1984, p. 271). Meanwhile, researchers and managers have recognized the importance of indoor environmental quality, which determines an employee's perceived work experience (Heinzerling et al., 2013) and includes office layout, furnishing, thermal, visual (lightning), and acoustic conditions, as well as indoor air quality (Frontczak & Wargocki, 2011; Kim & de Dear, 2012).

Frontczak and colleagues discovered in several studies that an individual's subjective perception of indoor air quality, also referred to as *perceived air quality* (von Kempski, 2002), is one of the most important factors influencing (work) satisfaction (besides thermal comfort) (Frontczak et al., 2012; Frontczak & Wargocki, 2011). Poor indoor air quality can lead to unspecific negative health effects – the so-called *sick building syndrome*[76] – lower satisfaction levels, and higher error rates, while good indoor air quality positively affects health, stress, productivity, well-being, and satisfaction of a building's occupants (Heinzerling et al., 2013; Jones, 1999; Kim & de Dear, 2012; Twardella et al., 2012; von Kempski, 2002; Zalejska-Jonsson & Wilhelmsson, 2013).

In this context, von Kempski (2002, 2004) suggests that indoor air must not only be objectively unpolluted (i.e., hygienically clean), but also perceived as subjectively pleasant, natural, and fresh. One possibility to enhance indoor air quality

---

[76] Sick building syndrome (SBS) describes "a constellation of symptoms that have no clear etiology and are attributed to exposure to a particular building environment" (Wang et al., 2008, p. 114). For a general overview, see Norbäck (2009); for a discussion in the context of ambient scents, see A. Girard (2015).

is to diffuse ambient scents to create pleasant, fresh breathing air (Girard et al., 2015; von Kempski, 2002, 2004): "Acceptable indoor air quality can only be achieved if the reduction in air pollution is combined with the addition of natural olfactory stimulants with sufficient positive attributes" (von Kempski, 2002, p. 61).

Taking these findings into consideration, we need to ensure in our empirical study that participating employees evaluate the ambient scent stimulus as pleasant in order to achieve positive responses (von Kempski, 2002).

3.2 The Effects of Ambient Scents in Work-related Contexts

So far, to our best knowledge, there are no empirical studies on the effects of ambient scents on employees in their actual workplace. However, in a recent literature overview, Girard et al. (2015) document that ambient scents applied in laboratory experiments in work-related contexts can affect individual's emotions, cognitions, and behavior.

The published findings are not consistent regarding whether or not there might be any effect of pleasant ambient scents on employees' mood: Some studies find a positive influence (Baron, 1990; Baron & Thomley, 1994; Knasko, 1995), while others find no effect (Gilbert et al., 1997; Knasko, 1993; Ludvigson & Rottman, 1989). However, unpleasant scents seem to lead to negative emotions (Asmus & Bell, 1999; Knasko, 1992; Rotton, 1983). Therefore, we conclude that avoiding unpleasant odors in the workplace should be critical for managers in general, as unpleasant scents also increase employees' motivation to escape the environment (Asmus & Bell, 1999).

In line with the findings of indoor air quality research, pleasurable ambient scents seem to enhance the evaluation of the environment (Baron, 1990; Baron & Bronfen, 1994), and reduce stress levels in work-related laboratory scenarios (Baron & Bronfen, 1994; Baron & Thomley, 1994). Furthermore, a pleasant ambient scent enhances individual goal and performance assessments in stimulated work contexts (Baron, 1990). However, individuals report negative effects on health and performance under the influence of unpleasant scents, although no such actual effects were recorded (Knasko, 1993).

Moreover, certain olfactory stimuli can enhance individual performance in routine tasks and jobs (Barker et al., 2003; Ho & Spence, 2005; Raudenbush et al., 2002; Sakamoto et al., 2005; Warm et al., 1991), mathematics-based operations (Degel & Köster, 1999; Gilbert et al., 1997; Knasko, 1993; Ludvigson & Rottman, 1989), and language-based tasks (Baron & Bronfen, 1994; Baron & Thomley, 1994; Degel & Köster, 1999; Herrmann et al., 2013; Knasko, 1993; Rotton, 1983), while unpleasant scents distract employees lowering responsiveness and performance and increasing error rates (Habel et al., 2007; Nordin et al., 2013; Rotton, 1983). However, if such effects "will dissipate or continue with longer or repeated exposures is unclear and also requires further research" (Gaygen & Hedge, 2008, p. 90).

In sum, the reviewed studies from work-related laboratory experiments suggest that unpleasant scents in the workplace will negatively influence exposed individuals, while pleasant ambient scents predominantly trigger positive responses. We therefore argue that the olfactory situation generally plays a key role for employees in their workplace. Moreover, we found indication, that ambient scents can evoke (positive and negative) internal and behavioral responses in employees.

4   Qualitative Research Design

Based on previous findings, we know that the air quality in the workplace is a key aspect of satisfaction and well-being of employees. Moreover, laboratory studies indicate that ambient scents are able to induce a broad range of internal and behavioral responses. However, so far we do not know how pleasurable ambient scent diffused in the workplace actually influences employees. In order to further elaborate our research questions, we conducted a qualitative diary study, which we now present.

4.1   The Diary Methodology

Despite Lewin being an experimental empiricist, he claims that holistic descriptions of an environment (e.g., those by novelists) are most valuable to science

(K. Lewin, 1946; Lück, 1996).[77] Thus, as suggested by field theory, we followed a qualitative holistic research approach, and chose a description and reflection of the situation as a whole before analyzing specific elements of the environment (Hawkes et al., 2009; Kassarjian, 1973; K. Lewin, 1942, 1946): Data was collected using the diary methodology, which "refers to research in which participants report on experiences during their normal day-to-day life" (Simons & Parkinson, 2009, p. 176). The diary methodology is specifically appropriate for scent research, since it also captures small events in employees' life (work)space, considering all facets as relevant to the respondents (Wheeler & Reis, 1991). Furthermore, this methodology is suitable to investigate the following questions (Bolger et al., 2003): What is typical for a person? What are between-person differences and what are their sources? This is especially important in the context of scent research, since scent perception – in terms of scent acuity as well as scent preferences – differs between individuals (Girard et al., 2015; Gulas & Bloch, 1995). Therefore, our study also seeks to identify similarities between persons, and might conclude in different clusters of ambient scent effects on employees.

We chose a time-based design with a fixed schedule, where the participants personally had to fill in the diary with paper and pencil regularly and contemporaneously at the end of each workday (Alaszewski, 2006; Bolger et al., 2003), in order to avoid any recall bias due to a memory gap (Kahneman et al., 2004b; Simons & Parkinson, 2009). This methodology offers the advantage that participants can take as much time as they need to update their diary, can freely answer questions, can provide as much personal information as they want, and can express their topics' priorities without the pressure of an interview-like situation (reduction of social desirability). However, the methodology requires high participant cooperation and commitment, since it is fairly time-consuming (Bolger et al., 2003; Hawkes et al., 2009; Simons & Parkinson, 2009). This is why its use is limited to a good number of repetitions, in order to avoid habituation to the questions and reactance[78] (Bolger et al., 2003; Brooks, 1987).

---

[77] In interpretation of Lewin's work, Lück (1996) suggests that there is no need for quantitative research methods to deliver valuable results.
[78] Remarks concerning reactance see Footnote 16.

In general, a diary period of four to 10 days is recommended for qualitative studies, in order to avoid negative emotions on the part of participants (Brooks, 1987; Hawkes et al., 2009; Simons & Parkinson, 2009; Waddington, 2005). Therefore, we chose a diary design over five working days, including a description of participant's workday experience on two days with scent, one day without, and again two days with exposure to a specific ambient scent in their workplace (see Figure 8). For this purpose, we provided every participant with an own scent cartridge, which he had to open and place in her/his office autonomously on each scented day.

Figure 8: Overview of Setting Diary Study[79]

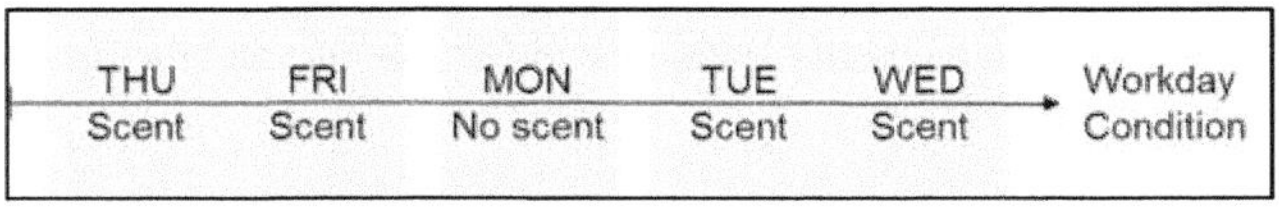

THU	FRI	MON	TUE	WED	Workday
Scent	Scent	No scent	Scent	Scent	Condition

Choosing this specific design allowed us to observe whether or not there were any variations in scent perception and/or scent influence over time, since diary studies are especially appropriate for research objectives that have a temporal structure, which might differ between as well as within individuals (Simons & Parkinson, 2009).

4.2   Detailed Description of the Diary

We designed the diary based on the recommendations of Kahneman et al. (2004b) as a written questionnaire with structured, open-ended, and closed questions. We also considered the guidelines of Bolger et al. (2003): We printed the diaries in an easily portable format, and sought to reduce the possibility of participant errors by preprinting the diary for scheduled dates, with an indication whether the specific day it is a scented day or not. Also, one individual of our relevant sample population proofread our diary upfront in order to avoid any difficulties; and finally we stayed in personal yet unobtrusive contact with all participants during the study, since personal contact is more valuable to creating commitment and to retaining participants than monetary incentives and goodwill towards research (Symon, 1998).

---

[79] Own illustration.

The body of the diary consisted of several *packets* (Kahneman et al., 2004a) and can be found in Appendix D.1.

Packet 1: Packet 1 included the introduction, demographics, scent acuity questions (e.g., smoking habits, chronicle nasal illnesses, and sensitivity to specific scents), personal scent preferences, and personal scent use (e.g., perfumes). The participants had to answer these questions only once, on the first day, while the questions of packets 2 to 4 were due for reporting every workday.

Packet 2: The second packet comprised general evaluations of the workday, mood as well as health status on the specific day.

Packet 3: In packet 3, participants had to identify single episodes of their workday as a general summary and to create transparency of the overall time spent in the office.

Packet 4: In the fourth packet, the participants had to first openly describe their emotions and impressions during the workday, followed by direct questions related to their motivation, satisfaction, stress, and perceived performance.[80] Additionally, we included a closed feelings/emotions scale, as suggested by Kahneman et al. (2004a) to be able to also indirectly measure potential scent effects and compare them with self-reported reactions. To obtain a holistic picture, we further asked for critical incidents and other olfactory stimuli (besides the ambient scent) present that day in the workplace.

Packet 5: Additionally, on each scent day, the participants had to report how they perceived the scenting, the effects of the ambient scent, and if there was any variation in scent awareness/perception during the day. On day 3 without ambient scent, they had to report how they felt without the stimulus.

Packet 6: Finally, on the last day, each participant had to evaluate the overall scent experience during the week. We also asked how they would feel about a future use of ambient scents in their workplace, what effects they expect on themselves as employees, and what requirements regarding the scent stimulus itself they want to be considered (e.g., scent type, intensity).

---

[80] We had identified these aspects as most relevant for employees based on the empirical insights from prior research (see section 3).

Overall, keeping the diary took about 10 to 20 minutes per day, depending on the level of detail provided by each participant.

4.3   Characteristics of the Participants and the Research Setting

The study sample consists of research assistants from several institutes of a German business school, who were rewarded with a cabaret voucher and a scented candle as tokens of appreciation for their participation. We personally invited 10 employees to participate in the study, and we received seven diaries at the end of the study period. Owing to the study's qualitative character, a small sample between five and 20 participants seems appropriate (Kuzel, 1999). Our sample is based on a purposeful selection of research assistants as typical office employees, males and females from different institutes. This sampling strategy is especially appropriate for biographical research methodologies such as diary research (Kuzel, 1999; Symon, 1998). We individually visited each respondent to hand over the diary and to explain the study requirements. We stated our interest in understanding their day-to-day work lives, and their perceptions of the ambient scent. We had the impression that the participants were happy to take part in our project as long as it did not interfere with their workday activities. Potentially, the study sample is biased by some underlying self-selection mechanisms, since as faculty colleagues, we knew all participants personally. However, since diary studies need a high participant commitment, we accepted this potential bias as part of our study design and, correspondingly, as a limitation (Symon, 1998).

We collected a total of 34 diary entries in summer 2011 (7 x 5 = 35; one day is missing for one participant). We surveyed two female and five male employees, with an average age of 28.3 years (between 26 and 31 years, median = 29). Three participants were smokers, while nobody reported suffering from any known chronicle nasal illness or scent sensitivity impairment. The employees worked between 25 and 58 hours during our study week (median = 42 hours), and had at least 12 months' work experience (median = 29 months) (see Table 5).

Table 5: Study Participant Characteristics[81]

No.	Gender	Age	Smoker	Working hours	Tenure (months)	Private scent use
1	Male	26	No	58	29	Yes
2	Male	29	Yes	50	45	Yes
3	Male	26	No	37	12	Yes
4	Male	28	No	56	17	No
5	Male	31	Yes	25	46	Yes
6	Female	29	Yes	42	27	Yes
7	Female	29	No	42	58	Yes

Six out of seven employees use scented products in their private lives, for instance perfume or scented candles (all except of male, 28). One employee even reported having used an ambient scent in his office prior to participating in our study (male, 31).

Five participants had a single office, while two shared their office with one other colleague. Overall, the work environment can be characterized as pleasant, since everybody had enough space, visual privacy, functional equipment and furniture, a fairly low noise level, a sufficiently clean workspace, and appropriate colors and textures (white walls, no carpet) (Kim & de Dear, 2012). Furthermore, the air quality was neutral without specific scents already present in the environment. However, owing to the fact that none of the offices were equipped with an air conditioning system, several participants perceived the indoor air as fairly 'sticky' on some days during our study. Overall, the offices' high environmental quality might be a reason why six participants reported being satisfied with their job, while one was even very satisfied.

4.4 Description of the Scent Stimulus

We chose a scent mixture made up of jasmine, melon, violet leaves, and rosewood created by a professional perfumer, as scent stimulus for our study. Since, in real-life situations, most scents are comprised of several components and mono-scents are said to be incapable to evoke 'natural' and non-artificial effects (Kirk-Smith & Booth, 1987; von Kempski, 2002), we believe a complex scent composition fits better to our research target of investigating ambient scent's impact on employees in a real-life work context (Girard et al., 2015).

---

[81] Own illustration.

The selected ambient scent was intended to promote inner peace as well as relaxation, and to mask any unpleasant odors that might be present in the environment. Due to the lack of a centralized air conditioning or ventilation system, the scent stimulus was diffused via a small scent cartridge provided to each participant at the start of the study. The participants had to independently open their scent cartridge on scent days and place it individually somewhere in their office space (see Appendix D.2).

Since the hedonic evaluation of the pleasantness of a scent stimulus is generally key for successful scent selection (Engen, 1972), we conducted a pretest with a written questionnaire of the scent's properties with 32 research assistants from various institutes of the same German business school who did not participate in our diary study. Of the pretest participants, 63% were male, aged between 24 and 40 (mean = 28.2 years), and mostly non-smokers (81%). Tenure varied between four and 60 months, with an average of 27.2 months.[82,] We asked the participants to sniff the scent presented in an identical cartridge as in our diary study, and to evaluate its (un)pleasantness as well as its arousal level (*relaxed/tense*) on a bipolar scale (-3/+3), as suggested by Fisher (1974) and Doucé et al. (2013). As expected, the participants rated the scent as pleasant (M = 1.31, σ = 1.55), and an one-sample t-test revealed that it was significantly different from the scale's midpoint of zero (t(31) = 4.78, p = .000, r = .651) (Field, 2013). Furthermore, the scent was perceived as slightly low arousing (M = -.06, σ = 1.41), while this effect did not significantly deviate from the neutral scale's midpoint of zero (t(31) = -.250, p = .804, r = .045). Finally, we asked the respondents whether or not they rate the scent as familiar (*The scent is familiar to me*) on a seven-point Likert scale (*fully disagree/fully agree*), as suggested by Morrin and Ratneshwar (2003), to avoid any influence and bias from previous scent experiences. The respondents rated the familiarity as moderate (M = 4.44, σ = 1.63), and not significantly different from the scale's midpoint of 4 (t(31) = 1.52, p = .138, r = .263).[83]

Overall, we already obtained a first indication that our scent stimulus is evaluated as pleasant by the relevant population of office employees. However, as the

---

[82] The retrospective pretest took place in July 2014 (Campbell & Stanley, 1966). The participants received no token of appreciation, since the questionnaire was very short.
[83] We conducted the quantitative analyses with the statistic software IBM SPSS Statistics 21.

pretest was just a snapshot, the diary analysis results will reveal whether or not this perception and evaluation will remain consistent over five workdays.

4.5 Content Analysis

To analyze the diary's open questions, we conducted a qualitative content analysis following the approach by Mayring (2000): Two coders classified the responses into a mainly inductive category schema, which emerged from an open-minded data analysis. However, as some aspects of the diary are based on preliminary theoretical considerations, which are reflected in the dairy structure, the category schema also has some deductive elements.

Based on the aim of our explorative study – to identify the effects of ambient scent on employees – both coders separately worked through two of the seven diaries progressively, and every aspect related to the research questions was either matched to an existing category, or a new category was formulated. After comparing the two classification schemas separately arranged by the coders and a revision of the categories, a final set of complete coding instructions was prepared with category definitions, prototypical text passages, and mutually aligned coding rules. The revised schema was pretested in a further step, with all seven diaries, and led to no further adoptions. Finally, two experienced and trained judges separately coded the data into the final exhaustive categories. In case of any deviating categorization, the judges mutually agreed on one coding after the final comparison of the results (Kassarjian, 1977; Krippendorff, 1980; Mayring, 2000; Perreault Jr. & Leigh, 1989).[84]

Finally, we entered the coding results into an Excel spreadsheet to facilitate the subsequent analyses. In total, we identified 10 main categories strongly related to our central research questions (see Table 6). The participant answers resulted in 409 statements classified in 85 subcategories.

In order to provide highly qualitative results, we considered several criteria: By following the recommendations of Kolbe and Burnett (1991) – we applied a rule-governed approach, used experienced and independent judges, and pretested our category schema – we sought to deliver highly objective results, achieving an objectivity index score of 4 out of 5 (see Appendix D.4). Both the percentage

---

[84] For a step model of category development see Appendix D.3.

of agreement (0.954) as well as the interjudge reliability (0.976) indicate a highly consistent and reliable allocation of the data to the category schema by the two judges (Perreault Jr. & Leigh, 1989). Since all statements are furthermore selectively and fully represented by the categories and subcategories of our classification, we assume a satisfactory content validity of our study (Keaveney, 1995). However, "[h]ow categories are defined (…) is an art" (Krippendorff, 1980, p. 76) and, therefore, there will always remain room for interpretation.

Table 6: Main Categories and Definitions[85]

No.	Main category	Definition
1	Private scent use	The participants use scents in their private life
2	Experiences in the workplace	Unsupportedly (without direct question) reported experiences in the workplace, including bodily state, mood, time pressure, sense of achievement, and further critical incidents
3	Evaluation of feelings in the workplace	Answers to direct questions related to the employee's motivation, work satisfaction, stress, and capability
4	Scent aspects during scent days	Category includes all aspects of the ambient scent use during the scent days: scent evaluation, association, perception, and its effects
5	Scent aspects during scent-free days	All aspects related to the scent on scent-free days as feelings when scent was not used
6	Evaluation of scent use during study	Overall evaluation of scent use during the study as positive, neutral, or negative
7	Attitude towards permanent scent use	Approval (with reservation) or rejection of permanent scent use
8	Requirements for ambient scent use	Further requirements for permanent scent use: subtle, pleasant, and interchangeable
9	Assumed scent effects on oneself	Assumption of ambient scent effects when permanently used: positive, negative, and no effect
10	Further comments	All further comments related to permanent scent use in future without being a requirement (category 8)

5   Results and Discussion

Before sharing the results of our diary study and discussing the answers regarding our initially outlined research questions, we will provide a short description of the employees' life (work)spaces in order to generate a holistic picture (Appendix D.2 provides some impressions of the scent use during the study). Selected quotes from the diaries serve as illustrations.[86]

---

[85] Own illustration.
[86] As the diaries were gathered in German, the quotations were translated by the author.

Overall, our employees evaluated their five workdays as "pretty typical", and they were mostly "in a mildly pleasant mood" during that week. The participants rated themselves as motivated most of the time (22 out of 32 entries; subsequently abbreviated as 22/32; two are missing for this evaluation). One employee stated that "the motivation to work is very high every day, as otherwise one cannot do this job" (male I, 26). The perceived stress levels differed considerably between workdays and people. However, the employees mostly rated their work performance as high (12/33) or average (16/33), and were generally satisfied with their job as research assistants (30/32).

Table 7 provides an overview of our results, which we will discuss in depth in the following sections.

Table 7: Overview of Results[87]

No.	Described as	Hedonic scent evaluation	Attitude towards scents in workplace	Scent perception during day	Habituation on scent	Self-reported scent effects	Actual observed effects	Future scent use in workplace
1	Male I, 26	Positive with limitations	N/A	Temporarily, decreasing, increasing, unconsciously	N/A	Calming and cheerful, none	Enhanced enjoyment of work and capability assessment	Acceptance with requirements
2	Male, 29	Positive with limitations	N/A	Temporarily	N/A	None	Reduced stress and capability assessment	Refusal
3	Male II, 26	Negative	Inappropriate	Temporarily, increasing, decreasing	Mentioned	Reduced concentration, headache, nervousness	Reduced stress	Refusal, would affect choice of employer
4	Male, 28	Neutral	N/A	Temporarily, increasing, unconsciously	N/A	None	Reduced stress and capability assessment	Refusal
5	Male, 31	Positive	N/A	Decreasing, temporarily	N/A	None	None	Acceptance
6	Female I, 29	Positive	N/A	Temporarily, decreasing, continuous	N/A	Cheerful, nervousness, none	Enhanced enjoyment of work, reduced stress and negative emotions	Acceptance with requirements
7	Female II, 29	Positive with limitations	Inappropriate	Temporarily, unconsciously	Mentioned	Reduced concentration, none	N/A	Acceptance with requirements

---

[87] Own illustration based on results of the content analysis. N/A = not available.

5.1   Did the Employees Evaluate the Utilized Ambient Scent as Pleasant?

First, we will analyze, whether or not the employees evaluated the utilized ambient scent as pleasant. Five out of our 7 respondents evaluated the chosen ambient scent as pleasant at least on one day, especially if other external unpleasant scents were present in their workplace. For instance, once, one employee had to babysit a puppy that was "not yet domesticated; so I was happy about the ambient scent" (male, 29).

Another employee evaluated the scent as not unpleasant, and stated that he almost didn't perceive it (male, 28). One employee evaluated the scent as unpleasant: The scent was "rather annoying, the whole day I had to keep a window open, as I got headache from the scent. I am a guy who prefers fresh air" (male II, 26). Furthermore, he and another employee evaluated the scent as incongruent to their workplace: "I don't associate my workplace with scent. It is rather unfamiliar and hence slightly disturbing" (male II, 26). The other participant stated that the scent "is not unpleasant, but still strange, somehow not fitting", despite the fact that she evaluated the scent per se as pleasant (female II, 29).

Interestingly, one participant reported that the evaluation of the scent was "slightly worse, because of a slightly bad mood" (male, 29). This might imply that mood as situational factor influences scent evaluation.

5.2   Did the Employees' Perceptions of the Ambient Scent Vary Over Time?

Next, we were interested if and how the employees' perception of the ambient scent varied over time. Seven out of 40 entries related to scent perception stated that the perception decreased over the day, which is in line with the expected effect of adaptation: "In the beginning, [the scent was] very intense, until I no longer perceived it" (male I, 26).

However, 6 out of 40 entries suggest that the scent perception increased over the day, which contradicts the adaptation assumption: "Initially positive and fresh. At some point in time, [the scent] somehow became annoying (in the afternoon)" (male, 28). As mentioned in the settings' characteristics, the study took place in a week with fairly high outside temperatures. As a result, 4 out of 7 employees rated the scent's intensity as too high: "The warmer it became during the day, the stronger the scent was" (male II, 26). "Partly, I felt like in a Tibetan brothel junk[88] – [the scent was] too sweet and too intense (…) because of the heat in the office" (male I, 26). Therefore, one employee suggested that "perhaps [employers] should use scent only in the morning (once), so that [it is] unremarkable and becomes less intense over the day" (male, 28).

Furthermore, as it was not possible to introduce the ambient scent into the whole building, the employees used the scent only in their office and this locally limited diffusion fostered conscious scent perception (15/40): "Every time I entered my office, I perceived the scent for a while" (male I, 26). One employee concluded: "As a result of scenting only one room, [the scent] again attracted attention every time. If all rooms were scented, one would not perceive it anymore" (male I, 26).

Other reasons for temporary conscious scent perception were related to directed attention to the scent (e.g., seeing the scent cartridge on the desk) (5/40), after airing the office (2/40), and during short breaks from concentrated work within the office (2/40). However, this temporal scent perception indicates that the respondents became unaware of the scent after a while, which militates for scent adaptation. In two cases, the respondents reported that they didn't perceive the scent at all, and one reported that he perceived it consciously without variation.

As we were also interested in the habituation effect of scent perception, we created the research setting over five workdays. Two participants provided indication for habituation: "Okay, slowly I get used to it" (female II, 29). Her colleague stated: "The scent starts to 'become part of the furniture'" (male II, 26). Interestingly, these two employees initially evaluated the scent as incongruent to their workplace. Perhaps they assimilated their existing workspace schema to the new experience with ambient scent over time (Mandler, 1982).

Overall, in line with field theory, we could show that, firstly, the environment affects scent perception, for instance, with increased room temperature the scent perception became more intense. But, secondly, adaptation and habituation

---

[88] In German: "Tibetanische Puff-Dschunke".

within individuals also affected how the ambient scent was perceived in the environment. In conclusion, our employees' scent perceptions and awareness did vary over time.

5.3 Did the Introduction of a Pleasant Ambient Scent in the Workplace Lead to Internal and Behavioral Responses?

To answer the question, whether or not the introduction of the pleasant ambient scent led to internal as well as behavioral responses, we chose a two-step approach.

In a first step, we directly asked the employees about their perceived reactions to the scent. The participants mostly reported that the ambient scent had no conscious effects on their emotions, cognitions, or behavior (15/28). For two employees, the scent had a "calming and cheerful [effect]. At least, the mood on this day was better" (male I, 26) (4/28). Nine out of 28 entries were related to negative responses, such as reduced concentration, headaches, and nervousness. Interestingly, almost all negative responses (8/9) were reported by the two employees, who evaluated ambient scents as generally inappropriate for their workplace (male II, 26 & female II, 29). It seems that the general attitude towards ambient scents in the workplace influences the self-reported effect of scents.

In a second step, we asked our respondents to rate their feelings on each day (*How did you feel during this workday?*) on a seven-point Likert scale (from *fully disagree* to *fully agree*), as suggested by Kahneman et al. (2004b), in order to indirectly identify potential effects of the ambient scent the employees might not be aware of. The dimensions of the scale comprised positive emotions (*happy, friendly*), negative emotions (*frustrated, depressed, angry, worried, criticized*), and performance assessment (*capable*). Furthermore, we adapted two items of the original scale to the context of work and created the dimension enjoyment of work (*enjoying myself at work, impatient for the day to end* (reverse-coded)). We also added *stress* as a relevant feeling in the workplace.

In order to get an indication of potential scent influence the respondents were not aware of, we calculated the average score for each dimension for every workday and then compared the mean of scent exposure days with the mean

on the day without scent for each employee. We considered it a difference between the scent and the scent-free days if the mean of the dimension equals at
least Δ1, and could not be ascribed to any external critical incident or negative
health status reported by the respective employee.[89] Thus, even for the interpretation of our quantitative data, we sought to understand the situation as a
whole, as suggested by field theory (K. Lewin, 1946).

Overall, our aim was to identify similarities as well as differences between the
seven employees: One employee (male I, 26) enjoyed his work more and felt
more capable during scented days than on the scent-free day. A group of three
employees rated lower perceived stress levels, but two of them also felt less
capable (male, 29 & male, 28). Interestingly, the third less stressed employee
was the one with a negative attitude towards ambient scents in the workplace,
who reported negative health effects (male II, 26). Perhaps, he attributed his
negative feelings to the scent and therefore it led to *annoyance* (Kelley &
Michela, 1980), which is "the complex of human reactions that occurs as a result of an immediate exposure to an ambient stressor (odour) that, once perceived, causes negative cognitive appraisal that requires a degree of coping"
(van Harreveld, 2001, p. 13).[90]

Another employee predominantly combined the positive aspects mentioned before: She reported higher enjoyment of her work accompanied by less negative
emotions, and lower perceived stress levels (female I, 29). For one employee,
we couldn't identify any unaware scent effect (male, 31), and for the last participant of our sample (female II, 29), we were unfortunately not able to draw any
conclusion owing to several negative external critical incidents and her poor
health status during the week (back pain).

Overall, we could identify positive employee responses to our chosen pleasant
ambient scent. However, if employees generally see ambient scent use in work
environments as inappropriate, negative effects might occur.

---

[89] The results remain unchanged, even if we use the median instead of the mean value as basis
for comparison, except for two additional effects: For male I, 26 we could identify higher stress
evaluation on scented days, and female I, 29 felt more capable on scented days.
[90] For a detailed discussion, see A. Girard (2015).

5.4 Is there a Difference Between Self-reported and Observed Scent Effects on Employees?

By comparing the results of directly vs. indirectly measured scent influence, we realized that for most of the employees (5/7), the self-reported effects and the identified unconscious scent effects diverge.

Three participants reported that they were not aware of any scent effects. However, we identified reduced stress levels accompanied by lower capability ratings for two of these participants (male, 29 & male, 28), and a reduced stress level as well as less negative emotions associated with higher work enjoyment levels for the third employee (female I, 29).

The employee with a general negative attitude towards scents in the workplace reported that the scent caused discomfort. However, we could not identify any negative (indirectly measured) scent effect. In fact, we even found indication of a lower perceived stress level (male II, 26). Another participant reported positive emotions due to the scent diffusion: We found indication for additional, unaware positive effects, such as enhanced enjoyment of work and higher capability (male I, 26).

Only for one employee were the ratings and actual effects consistent. He reported that the ambient scent had no effect, and we could not identify any scent effect via indirect measurement (male, 31). As already mentioned, we were unable to draw any conclusions for the last participant (female II, 29).

Generally, self-reported effects of ambient scent exposure and indirectly measured reactions seem to strongly diverge in employees.

5.5 Further Questions

At the last day of the study, we asked the employees several further questions:

First of all, we wanted to know what the employees think about a permanent use of ambient scent in their workplace. Three out of 7 employees were generally against permanent and omnipresent ambient scent diffusion in their office. One employee stated that "I can go without [ambient scent], even if it was temporarily pleasant" (male, 28). The other two evaluated the study's scenting as "less good", because they prefer fresh air in their workplace (male, 29 & male II,

26). The employee with the negative attitude even said that permanent scent use in his workplace "would critically influence my choice of employer" (male II, 26). In contrast, three other employees were in favor of permanent scent use in their workplace provided that "the scent is subtle and can be individually regulated and adjusted" (female II, 29). Their overall experience during our study was mixed however, since the scent diffusion and its associated intensity was not ideal, especially under high temperatures.

Second, we were interested in properties which ambient scents require in order to be accepted. Overall, the requirements for permanent acceptance of scent mentioned by those six employees were: A pleasant scent (4/13) that is "a bit more fresh and fruity than this one" (female I, 29), "less sweet, (…) not so unnatural" (male, 29), and used subtly (4/13) – "less intense and must not attract attention" (male, 28). Furthermore, it would be ideal if the scent is individually controllable (3/13), and situationally exchangeable (2/13): "Employees must be able to decide whether or not and which scent is used; [it] must be changeable daily" (female I, 29). This is especially important, since "scent perception is very subjective, and depends on the situation, and on your current state. Thus, a permanent scent use would disturb" (male II, 26).

One employee, who evaluated the scenting during the study as generally pleasant, would agree with permanent use of scent without any reservations, because his "office has a strange inherent smell" (male, 31). He had no further requirements and continued using the ambient scent even after the end of our study. We provided him with further scent cartridges upon specific request. Thus, we assume that the acceptance of ambient scents and their evaluation correlates with the presence of unpleasant smells in the workplace, as further two employees noted "there were no unpleasant smells. In case of stench, maybe I would have missed [the scent]" (male I, 26); "In my workplace there are no disturbing smells that need to be masked" (male II, 26).

Finally, we asked the participants about the effects they anticipate if scents would be permanently used in their office in the future. Five out of 7 respondents assumed that a continuous diffusion of scent might have a positive impact. A pleasant scent "is quite able to enhance well-being in the workplace" (female I, 29), and thus might create "permanently more feel-good factor in the office" (female II, 29). As a result, one employee concludes that if "one feels good (or better), this could enhance job satisfaction" (male, 28). Furthermore, an ambient scent could "mask bad smells, create a sense of cleanliness and neutrality" (male I, 26). For one employee, the scent had "no noticeable [effects and thus] definitely no long-term impact" (male, 29). Only the employee with the negative attitude towards ambient scents also assumed negative scent effects in case of permanent scent diffusion in his workplace (male II, 26).

6 Implications

Based on the theoretical considerations, previous empirical findings, and our qualitative results, we derive several implications for practitioners and academics, which we will present now.

6.1 Managerial Implications

Olfactory cues present in life (work)spaces do affect employees, for instance, a pleasant perfume worn by an employee, or stinking waste in a garbage bin. Today, indoor air quality mostly creates negative associations, and employees predominantly rate air quality as unsatisfactory (Frontczak & Wargocki, 2011). So, even if no ambient scents are diffused, poor indoor air quality will lead to negative reactions. Employers should be aware of the importance of perceived air quality to their workforce, and might thus consider air quality as an actively manageable company resource. Otherwise, "[t]he daily anticipation of going to work and possibly encountering ambient malodors, may take a toll on the morale of these employees" (Knasko, 1992, p. 34).

If employers decide to introduce ambient scent in order to improve the perceived air quality in their buildings, it should generally be evaluated as pleasant by the employees (rather than the employer). Employers therefore should know their employees' expectations concerning indoor air (von Kempski, 2002). Moreover, it is important that the scent's intensity is not too high. Hereof, companies must not only consider the ambient scent itself but also other aspects of the work environment: Especially thermal aspects – the most dominant factor in perceived environmental quality (Frontczak et al., 2012; Frontczak & Wargocki, 2011) – can increase or decrease the perceived intensity of scents and can thus influence individuals' scent responses. Hence, if a pleasant ambient scent is diffused subtly there might be positive effects on emotions and enjoyment at work, as well as reduced stress levels, and eventually even higher job satisfaction levels. Employees' capability assessments might be enhanced or reduced, depending on the individual. However, if employers with bad indoor air quality within their buildings remain idly, environmental quality research indicates negative effects on health and satisfaction. Here, pleasurable ambient scent might be at least a possibility to avoid or mitigate negative reactions.

Additionally, if an ambient scent is diffused, employees should be involved in the scent selection and implementation process, in order to ensure acceptance and to avoid negative reactions. Ideally, an employee is able to decide whether or not he or she will be exposed to scent in their workplace (von Kempski, 2004). If so, an employee should be able to choose her/his preferred scent and should be able to change it if needed. This approach would allow companies to benefit from the expected positive effect of pleasant scents on the majority of their staff with a general positive attitude towards scents, while avoiding a potential negative influence on the smaller group of employees not in favor of olfactory stimulation in their workplace. While such an individual handling might be possible in office buildings with separate rooms, it might be difficult for instance in open-plan offices. Also in service and retail companies, which nowadays mostly introduce ambient scents in order to influence their customers (Goldkuhl & Styvén, 2007) above-mentioned requirements are hardly realistic. This fact increases the importance of early involvement of employees during the creation or selection of ambient scents in business environments.

6.2   Limitations and Future Research

In our study, we provided first qualitative insights into the effects of ambient scents on employees. However, it is in the nature of qualitative research that the results are limited in terms of generalizability. Instead, we sought to provide a holistic description of the situation, so as to derive several questions for future research (see Table 8).

Table 8: Questions for Further Research[91]

Employer type
- Do employee reactions differ between various industries?
- Do employee reactions differ depending on the type of work (e.g., white vs. blue collar)?
- Do acceptance and effects of ambient scents correlate with the prevailing (negative) indoor air quality in the work environment?

Scent circumstances
- Are there different responses depending on the duration and/or frequency of scent exposure in the workplace?
- What effect does a scent, developed to please customers, have on employees, and vice versa? Is there a difference depending on the target group for the scent?
- Are there different responses depending on the evaluation of a scent as pleasant, neutral, or unpleasant?

Employee characteristics
- Does the general attitude towards scents influence responses to ambient scents?
- Are there different responses to ambient scent depending on whether or not employees were informed about its use upfront?

Internal and behavioral responses
- What are the differences in employees' reported vs. observed (e.g., via biological measurement techniques) scent effects?

Our sample included only seven participants, from whom we gathered 34 diary entries, which is however appropriate for a qualitative diary study (Kuzel, 1999; Symon, 1998). As previously mentioned, we knew all participants personally. Though, we see this fact rather as advantage than as limitation, since the participants were highly motivated to keep up their diary, and they shared very detailed and also very personal information about their work life. Future research should still investigate if our results can be verified in other industries, different types of work environments (e.g., white vs. blue collar), and with larger sample sizes.

As we chose an office building with a rather neutral air quality, we are not able to determine if the results hold true in a setting, where pleasant or unpleasant olfactory cues are already present in the work environment. Hence, further studies could investigate ambient scents influence in other industry settings with various indoor air conditions. It would as well be interesting to understand if the acceptance of ambient scents correlates with the existence of unpleasant scents in a work environment.

As the offices in our study had no air conditioning system, we were unable to keep the temperature and humidity stable. However, thermal aspects play an important role for environmental quality perceptions, particularly for indoor air

---

[91] Own illustration.

quality (Kim & de Dear, 2012). In our study, the scent perceptions intensified with the increase in temperature, with the result that employees' acceptance levels decreased (von Kempski, 2002). Thus, future research should try to introduce ambient scent via air conditioning or ventilation systems, which can ensure a consistent diffusion and distribution without intensity variations and are able to keep temperature and humidity controlled and stable.

With our diary design describing the whole workday, we were able to show that scent perception varied over the day in terms of adaptation. However, with only five workdays, the period of introspection is somewhat limited. We also found indications that habituation effects occurred owing to repeated exposure to the ambient scent, especially for employees who are not in favor of ambient scents in the workplace. The interesting question is whether or not our identified positive and negative scent responses and clusters persist over a longer time period. What happens if employees are exposed to an ambient scent repeatedly every workday over, for instance, a year? Do such effects persist or wear off after a while? Results from environmental quality literature imply that the influence of negative indoor air is persistent and employees will never get used to (Kim & de Dear, 2012; Norbäck, 2009). Thus, a longitudinal investigation with an employee panel could be very insightful.

For this study, the ambient scent and its distribution was selected by the research team and not purposefully developed and designed for the specific work environment and its employees. This might have led to less-than-ideal scent intensity on some hot summer days during our study, and there might have been a more suitable ambient scent for this specific work environment. However, considering scent selection, it would be interesting if there is a difference in employees' scent evaluation and more importantly scent effects, depending on whether the scent was developed to please and influence customers or employees – and vice versa. In other words, is there a difference depending on the target group the scent is actually intended for?

In our pretest as well as the main study, the employees predominantly evaluated the chosen ambient scent as pleasant. But even for those who evaluated the scent as rather unpleasant or neutral, we found indication for positive scent influence. Hence, it would be interesting to understand if there are any differences in ambient scent responses depending on an individual's evaluation of the scent as pleasant, neutral, or unpleasant.

In our study, at least one employee might have attributed negative health effects to his negative feelings towards ambient scent in the workplace: He reported negative effects, although we could observe only positive changes on scented days (indirectly measured). Thus, future research should verify that the reported influence of ambient scent is not only attributed to the presence of a scent per se. It would be interesting, for instance, to understand whether or not the reported responses to an ambient scent depend on an employee's attitude towards scenting in general, or if employees were informed about the scent use upfront. Laboratory studies suggest that even if no ambient scent is actually present, people might react negatively if they think the room was scented (e.g., Gilbert et al., 1997). These questions offer potential for various experiments, for instance, to compare self-reported scent effects and biologically measured scent influence for example via fMRI techniques, when considering respondents' general attitude towards ambient scents.

Overall, we call for future research in this in our view a completely under-researched area.

7 Conclusion

We know from environmental quality research that poor indoor air quality leads to negative health effects and lower job satisfaction levels. Therefore, it should be one of the first priorities of managers to ensure at least a relatively neutral air quality in the workplace in order to avoid negative reactions, for instance, via diffusion of pleasurable ambient scents. However, to date, research on the effects of ambient scents on employees is still in its infancy. Thus, research and experience will show whether or not pleasurable ambient scents are able to promote positive effects on, for instance, enjoyment at work, emotions, and job satisfaction. Even positive health effects might be possible, since olfactory cues have been used for medical purposes for centuries. Hence, the potential – positive or negative – influence of ambient scents on employees in their workplace should be taken into consideration by managers. Companies that only use ambient scent for marketing purposes to influence customers and increase sales should be aware that this practice might lead to *odor annoyance* on the part of their employees, and thus "a step backwards in the effort to improve the quality of indoor air" (von Kempski, 2002, p. 63). Such practices should therefore be discussed with affected employees beforehand in order to ensure involvement and acceptance.

8 Appendix Chapter D

Appendix D.1: Extract of the Diary (in German)[92]

Packet 1: Only on day one (introduction)

Liebe/r Teilnehmer/in,

Herzlichen Dank für Ihre Teilnahme an unserer Befragung!

Dieses Forschungsprojekt des Instituts für Marketing der Ludwig-Maximilians-Universität München beschäftigt sich mit der Wahrnehmung Ihres Arbeitsplatzes und Ihres Arbeitsalltags.

Ziel der Untersuchung ist es, Ihre alltäglichen **Eindrücke und Erlebnisse am Arbeitsplatz zu dokumentieren und besser zu verstehen.** Insbesondere geht es darum, festzustellen, wie Sie als Mitarbeiter einen **Dufteinsatz** am Arbeitsplatz wahrnehmen und empfinden.

Bitte nehmen Sie sich so viel Zeit wie nötig und beantworten Sie alle Fragen der Reihe nach. Da es um Ihre persönliche Wahrnehmung und Meinung geht, gibt es keine „richtigen" oder „falschen" Antworten! Beantworten Sie bitte alle Fragen vollständig und so ausführlich wie möglich.

Sämtliche Angaben werden selbstverständlich **absolut anonym** ausgewertet, so dass im Nachgang kein Rückschluss auf Ihre Person möglich ist. Die Daten werden ausschließlich im Rahmen des Forschungsprojekts ausgewertet und dienen wissenschaftlichen Zwecken.

Vielen Dank für Ihre Teilnahme – Sie leisten damit einen wertvollen Beitrag zu unserer Forschung!

Falls Sie Fragen haben können Sie sich gerne an Frau Anna Multani (multani@bwl.lmu.de / 089-2180-5738) wenden.

Bitte weiterblättern →

2

---

[92] Own illustration.

Packet 1: Only on day one (introduction)

Anleitung: Auf dieser Seite finden Sie eine Anleitung für die Verwendung Ihres Tagebuchs.

Als Teilnehmer des Forschungsprojekts erhalten Sie:

- eine tragbare **Duftkartusche**
- eine **Einwegkamera**
- sowie dieses **Tagebuch**

Bitte öffnen Sie an jedem Arbeitstag den Deckel Ihrer tragbaren **Duftkartusche** und platzieren Sie diese in Ihrem Büro (z.B. in der Nähe des Fensters). Falls Ihnen die Intensität des Duftes zu hoch sein sollte können Sie die Kartusche auch am Boden platzieren oder den Deckel nur halb öffnen. Am Ende des Arbeitstages verschließen Sie die Duftkartusche bitte wieder.

Mit Ihrer **Einwegkamera** sollten Sie zu Beginn jedes Arbeitstages die Position der Duftkartusche in Ihrem Büro festhalten. Außerdem dokumentieren Sie bitte besondere Vorkommnisse oder Eindrücke, die Ihnen im Verlauf des Arbeitstages wichtig erscheinen mit Ihrer Kamera.

Ihr **Tagebuch** dient schließlich dazu, Ihre Eindrücke des gesamten Tages am Abend zu beschreiben. Bitte füllen Sie Ihr Tagebuch daher **an jedem Tag der Studie zu Hause** aus! Das Ausfüllen wird täglich ca. 10-15 Minuten Ihrer Zeit in Anspruch nehmen. Sie finden die hierzu notwendigen Anweisungen auf den folgenden Seiten. Falls Sie es an einem Tag vergessen sollten tragen Sie den Tag bitte nach und vermerken dies.

Nach Abschluss der Studie bitten wir Sie die kompletten Unterlagen (Tagebuch, Kamera und Duftkartusche) am Donnerstag, den 4. August bei Frau Multani (Ludwigstr. 28 RG III Zimmer 322) abzugeben.

Vielen Dank und viel Erfolg bei Ihrer Teilnahme!

bitte weiterblättern →

Packet 1: Only on day one (demographics, scent acuity questions, scent preferences, and scent use)

Bitte füllen Sie diese Seite einmalig am ersten Tag der Studie aus (nur Donnerstag, 28.07.2011):

Zunächst haben wir einige allgemeine Fragen zu Ihrer Person:

1. Alter: _____

2. Geschlecht: ○ weiblich   ○ männlich

3. Wissenschaftlicher Mitarbeiter seit: _____

Nun haben wir einige allgemeine Fragen zu Ihrem Leben.

Bitte kreuzen Sie bei jeder Frage diejenige Antwortalternative an, die Ihre Meinung am besten widerspiegelt.

4. Alles in allem, wie zufrieden sind Sie mit Ihrem derzeitigen Job?

○ sehr zufrieden   ○ zufrieden   ○ nicht sehr zufrieden   ○ überhaupt nicht zufrieden

5. Wie zufrieden sind Sie derzeit mit Ihrem Leben generell?

○ sehr zufrieden   ○ zufrieden   ○ nicht sehr zufrieden   ○ überhaupt nicht zufrieden

6. Welche Faktoren tragen ganz Allgemein zu Ihrer Arbeitsplatzzufriedenheit bei?

_____
_____

7. Setzen Sie Düfte auch in Ihrem privaten Umfeld ein (bspw. Parfum, Duftkerzen etc...)? Wenn ja, welche und in welchem Ausmaß?

_____
_____

Da bei unserer Studie auch Ihre Geruchswahrnehmung relevant ist, würde uns noch interessieren:

8. Sind Sie Raucher?
   ○ Nein   ○ Ja

9. Leiden Sie an chronischen Erkrankungen der Atemwege?
   ○ Nein   ○ Ja

10. Haben Sie Duftempfindlichkeiten bzw. leiden Sie an Allergien in Bezug auf bestimmte Düfte?
    ○ Nein   ○ Ja, welche Düfte:_____

Bitte weiterblättern →

4

Packet 2: Every day (evaluations of workday, mood, and health)

LMU

Bitte füllen Sie diesen Teil des Fragebogens täglich am Ende Ihres Arbeitstages zu Hause aus.

Tag 1: Donnerstag, 28.07.2011

Waren Sie heute am Arbeitsplatz?

11.   O Ja   O Nein

Falls nicht, füllen Sie den heutigen Tag bitte nicht aus und fahren Sie morgen auf Seite 10 fort.

Zunächst würden wir gerne erfahren wie Sie sich am heutigen Arbeitstag insgesamt gefühlt haben.

12. Wenn Sie über den heutigen Tag nachdenken, welchen Anteil des Tages waren Sie...

in sehr schlechter Stimmung	_____ %
in gereizter Stimmung	_____ %
in verhalten guter Laune	_____ %
in sehr guter Laune	_____ %

Bitte verteilen Sie in Summe   100 %

Nun würden wir gerne erfahren wie typisch der heutige Arbeitstag für einen Donnerstag war.

13. Im Vergleich zu einem gewöhnlichen Arbeitstag, war der heutige Tag... (Bitte kreuzen Sie bei jeder Frage diejenige Antwortalternative an, die Ihre Meinung am besten widerspiegelt.)

Viel schlimmer	Etwas schlimmer	Ziemlich typisch	Etwas besser	Viel besser
O	O	O	O	O

Wie würden Sie generell Ihren heutigen Gesundheitszustand einschätzen?

14. Litten Sie heute an Erkältung, Schnupfen, Kopfschmerzen, etc.?

_____

_____

Bitte weiterblättern →

5

Packet 3: Every day (concrete episodes)

LMU

Waren Sie heute die meiste Zeit des Arbeitstages in Ihrem Büro oder hatten Sie z.B. Lehrveranstaltungen oder Sitzungen in anderen Zimmern?

15. Größtenteils im Büro   Häufiger Raumwechsel
     O   O

Bitte geben Sie uns einen kurzen Überblick über Ihre Verweildauer in Ihrem Büro bzw. in anderen Räumen:

	Raum	Anfangszeit	Endzeit	Besonderheiten
1				
2				
3				
4				
5				
6				
7				

Bitte weiterblättern →

Packet 4: Every day (emotions and impressions during workday)

Bitte denken Sie nun noch einmal über Ihren heutigen Arbeitstag nach.

16. Bitte beschreiben Sie, wie Sie sich während dieses Arbeitstages gefühlt haben.

    Gehen Sie dabei bitte insbesondere auf Ihre Stimmungen, Gefühle und Eindrücke ein:

    _____

    _____

    _____

    _____

    _____

    _____

    _____

    _____

    _____

    _____

17. Wie würden Sie Ihren Tag hinsichtlich folgender Kategorien beschreiben?

    a. Ihre heutige Arbeitsmotivation:

    _____

    _____

    b. Ihre heutige Arbeitszufriedenheit bzw. Arbeitsplatzzufriedenheit:

    _____

    _____

    c. Ihr heutiges Stresslevel während der Arbeit:

    _____

    _____

    d. Ihre heutige Arbeitsleistung:

    _____

    _____

    _____

Bitte weiterblättern →

Packet 4: Every day (feelings scale, critical incidents, and other olfactory stimuli)

LMU

Bitte beantworten Sie folgende Fragen nach dem Ausmaß Ihrer Zustimmung zu den einzelnen Aussagen. Zur Bewertung der Aussagen steht Ihnen eine 7-stufige Skala zur Verfügung, auf der Sie Ihre Zustimmung von „stimme überhaupt nicht zu" (linkes Ende) bis „stimme voll und ganz zu" (rechtes Ende) angeben und abstufen können. Bitte kreuzen Sie diejenige Alternative an, die Ihre Meinung am besten widerspiegelt.

Wie haben Sie sich während des heutigen Tages gefühlt?

stimme überhaupt nicht zu   teils-teils   stimme voll und ganz zu

	1	2	3	4	5	6	7
Ungeduldig dass der Tag endlich zu Ende geht	o	o	o	o	o	o	o
Glücklich	o	o	o	o	o	o	o
Frustriert	o	o	o	o	o	o	o
Niedergeschlagen	o	o	o	o	o	o	o
Leistungsfähig	o	o	o	o	o	o	o
Gehetzt	o	o	o	o	o	o	o
Freundlich	o	o	o	o	o	o	o
Verärgert	o	o	o	o	o	o	o
Besorgt	o	o	o	o	o	o	o
Freude an der Arbeit	o	o	o	o	o	o	o
Kritisiert	o	o	o	o	o	o	o
Müde	o	o	o	o	o	o	o

18. Gab es während des Arbeitstages besondere Vorkommnisse?

_____

_____

_____

_____

_____

19. Sind Ihnen heute besondere Gerüche am Arbeitsplatz aufgefallen?

_____

_____

_____

_____

_____

Bitte weiterblättern →

5

Packet 5: On scent days (perception and effects of the ambient scent)

20. Wie haben Sie den Duft in Ihrem Büro heute empfunden?

_____
_____
_____
_____
_____

21. Welche Wirkung hatte der Duft auf Sie?

_____
_____
_____
_____
_____

22. Haben Sie den Duft heute bewusst wahrgenommen? Falls ja, gehen Sie bitte insbesondere darauf ein, in welchen Situationen Sie den Duft bewusst wahrgenommen haben und wann eher nicht?

_____
_____
_____
_____
_____

23. Haben Sie den Duft heute über den Tag gesehen gleich wahrgenommen? Bitte beschreiben Sie die Duftwahrnehmung im Zeitverlauf.

_____
_____
_____

24. Vermissen Sie den Duft jetzt am Ende des Arbeitstages?

_____

25. Abschließend würden wir gerne noch erfahren, wo Sie die Duftkartusche heute platziert hatten?

_____

Vielen Dank. Sie sind am Ende der heutigen Befragung angekommen. Bitte fahren Sie **morgen** auf der nächsten **Seite 10** fort.

Bitte weiterblättern →

9

Packet 5: On the day without scent (feelings without scent)

LMU

50. Wie haben Sie es empfunden heute ohne den Dufteinsatz zu arbeiten?

_____

_____

_____

_____

51. Haben Sie den Duft vermisst? Wenn ja, in welchen Situationen?

_____

_____

_____

_____

Vielen Dank. Sie sind am Ende der heutigen Befragung angekommen. Bitte fahren Sie **morgen** auf der nächsten **Seite 20** fort.

Am Dienstag den 02.08.2011 verwenden Sie bitte wieder Ihre Duftkartusche!

Bitte weiterblättern →

19

Packet 6: Only on the last day (overall evaluation of scent experience)

LMU | LUDWIG MAXIMILIANS UNIVERSITÄT MÜNCHEN | FAKULTÄT FÜR BETRIEBSWIRTSCHAFT MUNICH SCHOOL OF MANAGEMENT

Bitte füllen Sie diese Seite einmalig am letzten Tag der Studie aus (nur Mittwoch, 03.08.2011):

82. Wie hat Ihnen der Einsatz des Duftes in der vergangenen Woche gefallen?

_____

_____

_____

83. Wie würde es Ihnen gefallen, wenn an Ihrem Arbeitsplatz in Zukunft dauerhaft Duft eingesetzt würde?

_____

_____

_____

84. Welche Anforderungen an einen Duft hätten Sie, damit Sie diesen akzeptieren würden?

_____

_____

_____

85. Welche Wirkungen hat ein solcher Dufteinsatz Ihrer Meinung nach auf Sie, als Mitarbeiter?

_____

_____

_____

86. Gibt es zum Thema „Duft am Arbeitsplatz" noch wichtige Aspekte, die aus Ihrer Sicht entscheidend sind oder die Sie uns gerne noch mitteilen möchten?

_____

_____

_____

Abschließend haben wir noch einmal zwei allgemeine Fragen zu Ihrem Leben.

87. Alles in allem, wie zufrieden sind Sie mit Ihrem derzeitigen Job?

O sehr zufrieden   O zufrieden   O nicht sehr zufrieden   O überhaupt nicht zufrieden

88. Wie zufrieden sind Sie derzeit mit Ihrem Leben generell?

O sehr zufrieden   O zufrieden   O nicht sehr zufrieden   O überhaupt nicht zufrieden

---

HERZLICHEN DANK FÜR IHRE TEILNAHME

Bitte übergeben Sie alle Unterlagen – das ausgefüllte Tagebuch, die Einwegkamera sowie die Duftkartusche – morgen an Frau Multani (Ludwigstr. 28 RG III Zimmer 322).

30

Appendix D.2: Impressions of the Scent Use during the Diary Study[93]

---

[93] Own illustration.

Appendix D.3: Category Development[94]

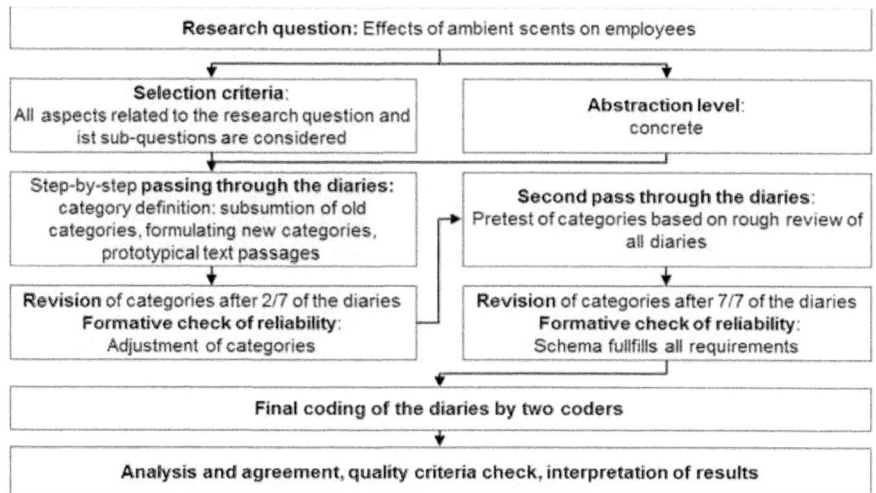

Appendix D.4: Objectivity Index[95]

Objectivity	Implementation in study	Dichotomous manifesta-tion (1 = yes, 0 = no)
Rules and procedures	Category schema with category defini-tions, prototypical text passages, and coding rules	1
Judge training	Category schema was developed by both coders; they were familiar with the sche-ma and its operational definitions	1
Measure pretesting	Category schema was pretested with a rough overview of the data material in order to ensure completeness	1
Judge independence	Autonomous judgments, however, one of the judges was the researcher	0
Number of judges > 1	Two coders	1
Sum		4

---

[95] Own illustration based on Kolbe and Burnett (1991).

E. GENERAL DISCUSSION AND CONCLUSION

This present thesis sought to contribute to current scent marketing research, especially on the effects of ambient scents on customers and employees. As each paper contains a separate discussion section and puts forth a set of specific implications, this chapter will summarize and conclude with a discussion of the main findings across all three papers, draw a set of overarching implications, and end with a final conclusion.

Prior to that, we now present a general discussion of measures taken to safeguard good scientific practice in the course of this dissertation project. Because the work in hand involved research on humans, this seems particularly relevant to the author.

1 Safeguarding Good Scientific Practice

First, the present dissertation consistently considered all general recommendations of the commission on professional self-regulation in science of the German Research Foundation (Deutsche Forschungsgemeinschaft, 2013). Second, we also took into account the specific requirements for research on humans claimed by the German Research Foundation (Deutsche Forschungsgemeinschaft, 2012).

Key elements of safeguarding good scientific practice generally include following professional standards and documenting all results, to foster transparency and objectivity. Each paper of this thesis comprises a detailed specification of the research approach, as well as a description and discussion of the findings. All primary data are securely stored in a durable form and can be examined upon request. Furthermore, during the research process, all results were continuously and consistently questioned by the author and other researchers, who are separately acknowledged for every article.[96] None of the co-authors received honorary authorship; they significantly contributed to this article with different shares.

---

[96] As Heinz Maier-Leibnitz (physicist and former president of the German Research Foundation) remarks: "Scientists are educated by their work to doubt everything that they do and find out ... especially what is close to their heart" (Deutsche Forschungsgemeinschaft, 2013, p. 92).

Moreover, as the empirical studies discussed in papers 2 and 3 involve research on humans (field experiment and diary study), further recommendations of the German Research Foundation regarding legal aspects, potential risks, test person selection criteria, and method of informed consent including ethical aspects, were considered, and will be discussed in the following (Deutsche Forschungsgemeinschaft, 2012).

First, to our best knowledge, the ambient scent used in this dissertation project meets all relevant German and European legal requirements, which seek to ensure that generally no hazardous substances or components are applied in the industry, and that all ingredients of a specific scent are declared in a material safety sheet (for our selected scent, see SCENTCOMMUNICATION, 2007).[97] Besides, the provider of the selected ambient scent stimulus submits itself to the stricter self-regulations of the International Fragrance Association. All these regulations seek to protect exposed people and the environment (Bundesministerium für Umwelt, 2005; A. Girard, 2015; International Fragrance Association, 2006; Umweltbundesamt, 2006a).

Furthermore, the diffusion of an ambient scent must not harm any study participant; thus, potential risks should be minimized and precautious measures taken. To date, no actual harmful health effects of ambient scents have been scientifically documented (Stroh, 2005). Especially, there is no evidence that inhalative scent exposure leads to allergic reactions (Hatt, 2007; Umweltbundesamt, 2006b). However, as scents are known as the second most common reason for contact allergies (e.g., perfumes or deodorants) (Umweltbundesamt, 2006a, 2006b), the olfactory stimulus used in this research project was diffused via gel containers, which emit the olfactory molecules through the natural air current or the air conditioning system without any further carrier substances (e.g., alcohol). Nonetheless, respiratory tract irritations might generally still be possible within special target groups (de Lima et al., 2011; Mücke & Lemmen, 2010; Umweltbundesamt, 2006a; Wang et al., 2008).[98]

---

[97] Relevant laws, for instance, Lebensmittel- und Futtermittelgesetzbuch (LFGB), Chemikaliengesetz (ChemG), Gefahrstoffverordnung (GefahrstoffV), REACH-Regulation (European Union regulation concerning the registration, evaluation, authorisation & restriction of chemicals).

[98] For a detailed discussion, see A. Girard (2015).

However, neither in approval tests by the producer nor during our studies did our ambient scent cause any irritations or other negative health symptoms with exposed participants we are aware of, except for reported discomfort of two employees, who evaluated the scent use as inappropriate for their workplace.

Regarding the criteria of test person selection for paper 2 – **Are You on the Right Scent? Ambient Scents' Short- and Long-term Effects on Customers in a Servicescape** – we selected test persons, who were commuters on a specific train section of a German railway company (*Deutsche Bahn*) with a daily workday commute of at least 15 minutes one way. Potential participants were personally addressed in the trains of the specific track section, or via mail to all subscribers in the company's customer database. All potential test persons were approached without pre-selection and were finally chosen based on the above criteria. Every individual participated voluntarily and received a non-monetary compensation after completing their study participation.

The test persons were only informed that their participation in the study would help to enhance the public transportation company's service, but not about the intended introduction of the ambient scent into the train compartments. Not to inform participants about the actual background of a study is common practice in marketing experimentation research (Aaker et al., 2001). The reason is to prevent and minimize test persons' reactions attributed to the study's background, which might thus bias the results regarding the effects under investigation (so-called *artifacts*) (Schnell et al., 2005). In scent research, previous empirical studies confirm the existence of such effects (Gilbert et al., 1997; Knasko, 1992; Knasko et al., 1990). However, it is worth noting that, with our research design, not only the study participants, but also all other customers of the specific train section were exposed to the ambient scent. So, strictly speaking, they also unwittingly participated in the experiment, even though they were not panel participants.

To be able to answer potential questions of all customers related to the ambient scent or the study background, the official customer service center of the German railway company was briefed and fully involved from the outset of our project. After the finalization of the experiment, all participants received a written explanation of the study background, the ambient scent diffusion, and its scientific intention. We did not receive any complaints as reaction to the ex post information.[99]

Nonetheless, ethical considerations should be discussed: The customers on the specific track section had no possibility to avoid the scenting, and it could thus be seen as a manipulation attempt (Goris & Hutter, 2011; Lunardo, 2012; Lunardo & Mbengue, 2013). The study sought to investigate whether or not the ambient scent diffusion influences customers. To be truly manipulative, the persuasion technique (scent) would need to directly determine customers' behaviors (von Rosenstiel & Kirsch, 1996). However, based on current scientific knowledge, a systematic control of individuals' emotions, cognitions, and decisions through mere exposure to scents seems impossible (Hehn, 2007; Knoblich et al., 2003). Pleasant ambient scents might exert a positive influence on customers; "[h]owever, it should be added that this does not happen systematically" (Maille, 2006, p. 109). In conclusion, ambient scent use should not been seen as manipulation, but rather as an attempt to positively influence and persuade customers (von Rosenstiel & Kirsch, 1996). As Eric Spangenberg (Marketing Professor, Washington State University) concludes: "One could argue that it's nothing more insidious than pleasant music" (Klara, 2012).[100] However, as one customer notes, "I don't want to be permanently exposed to music or to scent, nor to anything else" (see chapter C, section 4.2).

Finally, for paper 3 – **The Impacts of Ambient Scents in the Workplace: A Qualitative Investigation** – the test person selection criteria required participants to be employed as university assistants at the Munich School of Management, covering both males and females from different institutes. As the diary study was time consuming and required much commitment, all test persons were purposefully selected by the author. Every individual participated voluntarily and received a non-monetary compensation after participating in the study.

The participants were fully informed about the study's intention to investigate how employees perceive ambient scent use in their workplace. They received

---

[99] As a reaction to a national television broadcast on the research project more than a year after the study (see Nischk (2013, 09.07.)), we received one email from a concerned customer fearing an expansion and permanent scent diffusion in the train compartments of the German railway company. The railway company's customer service center followed up directly on the complaint.
[100] For a detailed discussion, see A. Girard (2015).

their own scent cartridge and were able to individually regulate the scent intensity. They thus agreed – in full informed consent and without any ethical dubi-
ousness – to the study intention.

Overall, we sought to consider all recommendations for good scientific practice
in the present dissertation. To conclude, we now present a summary of our
main findings.

2   Discussion of Main Findings

The summary of the results is based on the 11 major insights of the conceptual
paper 1, where our empirical findings are integrated.

Insight 1: The importance of olfactory stimuli as external environmental cues
depends on the service type.

Ambient scent might have a greater influence in services with high extent of
credence qualities (e.g., medical treatment), services where the actual experience is key (e.g., a cinema), and services where the customer stays within the
servicescape for a longer period (e.g., a hotel). In paper 2, we demonstrated
that ambient scents did act as an additional environmental experience cue for a
transportation service, adding value to the situational experience every time a
customer reuses the offering.

Insight 2: Furthermore, other scent sources besides ambient scent already
present in an environment should be considered, because an environment's
indoor air quality always influences individuals.

Based on our theoretical and empirical findings we assume that the poorer the
indoor air quality, the higher the acceptance of pleasurable ambient scent diffusion among individuals, and the greater its influence. As a customer suggested
in paper 2, a pleasant ambient scent could "compensate for the bad odors" and
could ensure that the "reek within the train would be neutralized" (see chapter
C, section 4.2). Also, one employee noted "there were no unpleasant smells [in
my office]. In case of stench, maybe I would have missed [the scent]" (see
chapter D, section 5.5). The overall acceptance of permanent ambient scent
diffusion was high in both empirical studies among exposed customers as well
as employees. Most individuals expected a positive influence of ambient scents

and generally were in favor of such practices if the stimulus was non-hazardous, subtle, and pleasant.

Insight 3: Scent is a multidimensional cue, and its objective characteristics include a scent's intensity, complexity, and arousal potential. A scent's influence on a specific individual is thereby also determined by the duration and/or frequency of its exposure.

In our empirical studies, we were predominantly interested in the temporal structure of ambient scents' influences. We introduced a dynamic perspective of scent exposure, distinguishing between short-term, long-term, and aftereffects of scent influence. In paper 2, we concluded that a scent's long-term effect depends on the dependent variable type: When investigating constructs that measure a situational evaluation of a service, ambient scent had a positive influence and contributed to the situational evaluation every time; thus, scent exerted a continuous long-term influence on customers. However, considering constructs that measure long-term attitudes, which evolve over time based on repeated service interactions, expectations adapted towards the new stimulus pattern and the influence of the pleasurable ambient scent leveled out over time. Our findings also suggest that a scent's aftereffects persist for a while, even after the removal of the stimulus.

In paper 3, we could show that employees' scent perceptions varied over time, as adaptation and habituation affected how the ambient scent was perceived in the environment; employees' awareness of the scent disappeared over time: "In the beginning, [the scent was] very intense, until I no longer perceived it" (see chapter D, section 5.2). However, "[a]s a result of scenting only one room, [the scent] again attracted attention every time" an employee entered the scented room (see chapter D, section 5.2). Such locally limited scent diffusion opens the possibility to foster conscious scent perception and leverage an attentional capture effect as well as an increased information processing capability of further environmental stimuli (Girard et al., 2015; Gould & Martin, 2001; Grabenhorst et al., 2011). This might be beneficial to support individuals facing highly complex tasks or tasks that demand high concentration in a specific indoor environment (e.g., a doctor's examinations during her/his medical rounds).

Insight 4: The actual perception of a scent can be conscious or unconscious, while the internal responses evoked by that scent remain mainly independent from the consciousness level of exposure.

In our first empirical study, most customers did not consciously perceive the scent: "I thought, somehow it smells different in here, but then I thought, it's probably just in my imagination" (see chapter C, section 4.2). Nonetheless, we were able to prove various – short-term and long-term – reactions to the ambient scent. In our second empirical study, the employees were fully aware of the ambient scent use. Interestingly, employees' self-reported reactions to ambient scent exposure and indirectly measured effects diverged strongly. Thus, independent of the actual level of consciousness, we were able to identify ambient scent effects on exposed individuals in both empirical studies.

Insight 5: An individual's scent acuity, and thus the perception of a scent, is said to depend on various individual moderators, such as demographical, physiological, situational, and personality-related or experience-related factors.

Even though our empirical research was not focused on moderating effects, we controlled for a multitude of potential factors in the experimental study (paper 2). We found no evidence for any moderation of gender, age, and smoking habits on ambient scents' influences on customers, with the exception of the relationship between gender and behavioral intention.

Insight 6: Ambient scents have subjective characteristics, such as its quality, hedonic value (preferences), and perceived congruence, which are individually evaluated, when perceived.

Even though the quality assessments of the chosen ambient scent differed between individuals, it was predominantly rated as pleasant in both of our empirical studies. Furthermore, the olfactory stimulus was at least not evaluated as incongruent by customers to the scenting's objective of creating an enjoyable train ride experience (paper 2). In contrast, two employees rated the scent as "not unpleasant, but still strange, somehow not fitting [to the workplace]" (see chapter D, section 5.1). However, at the end of the study period, the employees became familiar with and got used to the scent in their office: "The scent starts to 'become part of the furniture'" (see chapter D, section 5.2). We conclude that scent pleasantness is generally more important than congruence, as the fit to a specific context can and will be learned via repeated exposure and adjustment of respective schema.[101]

Insight 7: An ambient scent's effects depend on further external environmental factors, such as other ambient stimuli, as well as thermal factors, such as temperature, air pressure, or humidity, which act as moderators and influence possible reactions to scents.

As demonstrated in paper 3, the environment did indeed affect individual scent perception, for instance, with increased room temperature in the office, employees' scent perception became more intense: "The warmer it became during the day, the stronger the scent was" (see chapter D, section 5.2). Thus, ambient scent diffusion in commercial contexts is only recommendable to managers if one is able to control or at least keep constant all other stimuli and thermal factors, in order to ensure consistent stimulation patterns.

Insight 8: Perceived air quality mediates ambient scents' influences on internal and behavioral reactions of exposed individuals.

As already discussed with regards to insight 2, external scent sources influence individuals' PAQ – especially unpleasant olfactory stimuli present do have a negative impact. In paper 2, we could show that the pleasurable scent diffusion significantly improved the evaluation of exposed customers' PAQ. However, the actual mediation relationship was not part of this dissertation's empirical research program, and we recommend further research to test this relationship.

Insight 9 & 10: Because customers and employees are part of the same environment, the same olfactory cues will always exert effects on both parties in the servicescape, triggering physiological, emotional, and cognitive responses. Depending on these initial internal reactions, scents can lead to approach, avoidance, or behavioral coping behavior.

In paper 2, we could show an enduring positive influence of pleasurable ambient scent on exposed customers' perceived service experience, service quality, and service value; as well as a temporary positive promotion of satisfaction and brand attitude. Furthermore, for female customers, we discovered an enduring

---

[101] Also, see Engen (1972) and Mandler (1982).

positive effect on behavioral intentions, while this effect was only temporary for men.

In paper 3, focusing on employees' reactions, we found indirectly measured indications of more positive and less negative emotions, reduced stress levels, higher work enjoyment, and diverging capability ratings. Two employees directly stated the scent had a "calming and cheerful [effect]. At least, the mood on this day was better" (see chapter D, section 5.3); however, another two employees reported discomfort on scented days.

Insight 11: Scents in servicescapes are also expected to affect the quality and quantity of social interactions between customers and employees in a service encounter (as well as among each other).

However, testing the influence of ambient scents on social interactions was not part of the empirical investigations of this research project, and we leave it open for future research.

3 Implications

Reviewing and summarizing all three papers of this thesis, we would like to highlight the most important implications for management and academia, as follows:

Implications for Management: Most importantly, managers should be aware that even if no ambient scent is actively diffused within an environment, the indoor air quality will still affect exposed individuals. Thus, managers should ensure at least a relatively neutral air quality, so as to avoid any negative impacts of the olfactory setting on customers and employees. If a pleasant ambient scent is used to enhance the perceived air quality, the scent should meet several requirements: It should be rated as pleasant and evaluated as not incongruent with the brand, the environment, and the managerial purpose. Both the intensity and arousal level should be appropriately tailored to the specific context and target group. Owing to this complexity, we recommend investing in a dedicated scent selection and pretesting process, in cooperation with a professional perfumer and/or scent marketer. In general, companies should not forget that ambient scent is one aspect among many in an overall marketing/branding strategy and should thus be carefully integrated (Ravn, 2007).

An appropriate scent can contribute to various positive effects in customers and employees. It can thus be used as a marketing tool in multiple ways: To enhance situational customer evaluations of a service, add a sensory dimension to a brand's schema and image, and/or positively influencing employees by creating a "more feel-good factor in the office" (see chapter D, section 5.5). Overall, no negative impact was identified in our empirical studies, despite reported discomfort by two employees, who were not in favor of introducing ambient scents in their workplace. To ensure acceptance, "[e]mployees must be able to decide whether or not and which scent is used". However, companies must be aware that scenting the workplace might, for a few employees, even "critically influence (...) [the] choice of employer" (see chapter D, section 5.5). We therefore recommend informing and involving both employees and customers as early as possible about ambient scent diffusion in any (service) environment.

In the long term, positive or at least neutral effects of continuous pleasant ambient scent diffusion are expected. Even if discontinuing scenting, we found no indication for negative aftereffects. Generally, companies might consider permanent introduction of scents into their environments. However, without the risk of negative aftereffects, a temporal or seasonal diffusion of ambient scents might also make sense. For instance, around Christmas only, seasonal scent diffusion could support the festive ambience within a servicescape. Also, in case of a limited budget, a company could decide to only temporarily harvest ambient scents' benefits. Another possibility to deploy the potential of the olfactory sense is to diffuse scents only locally, limited to specific servicescape or workplace areas, to foster conscious scent perception and awareness in key locations, for instance a hotel lobby or a showroom.

Overall, (ambient) scent marketing seems to be able to positively influence internal as well as external target groups.

Implications for Future Research: With the cumulative character of this dissertation, we sought to successively address questions for further scent research called for in previous work. Starting with paper 1, we developed an integrative framework that holistically describes the process of olfactory stimulation and its impacts on customers and employees in service environments, and outlined a conceptual research agenda for further empirical investigation. In paper

2, we primarily focused on analyzing ambient scent effects on customers in a longitudinal field research setting. With paper 3, we addressed ambient scent influences on employees in a qualitative field diary study over a week.

However, some aspects remain open for future research: Especially social interaction under the influence of ambient scent requires further investigation, including the aspects of service encounter between customers and employees, but also among employees, for instance in an office setting. Further unclear facets of olfactory influence in a marketing context include PAQ's mediating role, and most importantly, ambient scent's impact on financial company performance. Furthermore, additional investigations in other research settings or other industries might further strengthen the generalizability of our results.

Overall, to our best knowledge, this dissertation provides the first empirical investigations that assess the dynamic and temporal structure of ambient scents effects over time and smoothens the way for further longitudinal research.

4   Conclusion

Leveraging the positive influences of pleasurable olfactory stimuli is not a new concept. Scents have been used for medical purposes since ancient times. Also, burning incense, scented candles, or wearing perfumes have been used by individuals or institutions to create a pleasant atmosphere or mask unpleasant (body) odors for centuries (Ackerman, 1990).

Nowadays, companies also seek to profit from the benefits of our sense of smell. Scent marketing has been a trend for many years in various service areas and a variety of settings. However, managers rarely understand all underlying concepts and mechanisms of scent perception and processing – especially concerning scent effects over time.

Thus, this dissertation extends the predominant static view of scent influence after one-time exposure to a more dynamic longitudinal view of repeated scent exposure. By doing so, the results can be more easily transferred into real-life business settings, where customers and/or employees usually stay in an indoor environment for an extended period – and return mostly more than once.

Overall, the results of this thesis indicate that ambient scent affects individuals positively or neutrally in the long term, and that no negative aftereffects of scent exposure need to be expected.

> *"Smell is really transporting. Seeing, hearing, touching, tasting are just not as powerful as smelling if you want your whole being to go back for a second to something."*
>
> (Andy Warhol, 1975, p. 151)

F.   REFERENCES

Aaker, D. A., Kumar, V., & Day, G. S. (2001). *Marketing Research* (7. ed.). New York, NY u.a.: Wiley.

Ackerman, D. (1990). *A Natural History of the Senses*. New York, NY: Random House.

Alaszewski, A. (2006). *Using Diaries for Social Research*. London; Thousand Oaks, CA: SAGE Publications.

Aldwin, C. M., & Revenson, T. A. (1987). Does Coping Help? A Reexamination of the Relation Between Coping and Mental Health. *Journal of Personality and Social Psychology, 53*(2), 337-348.

Amabile, T. M., Conti, R., Coon, H., Lazenby, J., & Herron, M. (1996). Assessing the Work Environment for Creativity. *The Academy of Management Journal, 39*(5), 1154-1184.

Anderson, E. W., Fornell, C., & Lehmann, D. R. (1994). Customer Satisfaction, Market Share, and Profitability: Findings From Sweden. *Journal of Marketing, 58*(3), 53-66.

Asmus, C. L., & Bell, P. A. (1999). Effects of Environmental Odor and Coping Style on Negative Affect, Anger, Arousal, and Escape. *Journal of Applied Social Psychology, 29*(2), 245-260.

Ayabe-Kanamura, S., Schicker, I., Laska, M., Hudson, R., Distel, H., Kobayakawa, T., & Saito, S. (1998). Differences in Perception of Everyday Odors: a Japanese-German Cross-cultural Study. *Chemical Senses, 23*(1), 31-38.

Badia, P., Wesensten, N., Lammers, W., Culpepper, J., & Harsh, J. (1990). Responsiveness to Olfactory Stimuli Presented in Sleep. *Physiology & Behavior, 48*(1), 87-90.

Baeyens, F., Wrzesniewski, A., de Houwer, J., & Eelen, P. (1996). Toilet Rooms, Body Massages, and Smells: Two Field Studies on Human Evaluative Odor Conditioning. *Current Psychology, 15*(1), 77-96.

Bagozzi, R. P., Gopinath, M., & Nyer, P. U. (1999). The Role of Emotions in Marketing. *Journal of the Academy of Marketing Science, 27*(2), 184-206.

Bakeman, R. (2005). Recommended effect size statistics for repeated measures designs. *Behavior Research Methods, 37*(3), 379-384.

Baker, J., Parasuraman, A., Grewal, D., & Voss, G. B. (2002). The Influence of Multiple Store Environment Cues on Perceived Merchandise Value and Patronage Intentions. *Journal of Marketing, 66*(2), 120-141.

Barker, S., Grayhem, P., Koon, J., Perkins, J., Whalen, A., & Raudenbush, B. (2003). Improved Performance on Clerical Tasks Associated with Administration of Peppermint Odor. *Perceptual And Motor Skills, 97*(3), 1007-1010.

Baron, R. A. (1981). Olfaction and Human Social Behavior: Effects of a Pleasant Scent on Attraction and Social Perception. *Personality and Social Psychology Bulletin, 7*(4), 611-616.

Baron, R. A. (1983). "Sweet Smell of Success"? The Impact of Pleasant Artificial Scents on Evaluations of Job Applicants. *Journal of Applied Psychology, 68*(4), 709-713.

Baron, R. A. (1986). Self-Presentation in Job Interviews: When There Can Be "Too Much of a Good Thing". *Journal of Applied Social Psychology, 16*(1), 16-28.

Baron, R. A. (1990). Environmentally Induced Positive Affect: Its Impact on Self-Efficacy, Task Performance, Negotiation, and Conflict. *Journal of Applied Social Psychology, 20*(5), 368-384.

Baron, R. A. (1997). The Sweet Smell of... Helping: Effects of Pleasant Ambient Fragrance on Prosocial Behavior in Shopping Malls. *Personality and Social Psychology Bulletin, 23*(5), 498-503.

Baron, R. A., & Bronfen, M. I. (1994). A Whiff of Reality: Empirical Evidence Concerning the Effects of Pleasant Fragrances on Work-Related Behavior. *Journal of Applied Social Psychology, 24*(13), 1179-1203.

Baron, R. A., & Thomley, J. (1994). A WHIFF OF REALITY: Positive Affect as a Potential Mediator of the Effects of Pleasant Fragrances on Task Performance and Helping. *Environment and Behavior, 26*(6), 766-784.

Baron, R. M., & Kenny, D. A. (1986). The Moderator–Mediator Variable Distinction in Social Psychological Research: Conceptual, Strategic, and Statistical Considerations. *Journal of Personality and Social Psychology, 51*(6), 1173-1182.

Bawa, A., & Kansal, P. (2008). Cognitive Dissonance and the Marketing of Services: Some Issues. *Journal of Services Research, 8*(2), 31-51.

Bayrisches Landesamt für Statistik und Datenverabeitung. (2013). Erwerbstätige sowie Schüler und Studierende nach Pendlereigenschaften in Bayern 2012 – Ergebnisse der 1%-Mikrozensuserhebung 2012 (Vol. A VI 2/S4 4j 2012). Munich.

Bell, S. (2007). Future sense: defining brands through scent. *Market Leader, Autumn/38*, 60-62.

Berlyne, D. E. (1971). *Aesthetics and Psychobiology* (1. ed.). New York, NY: Appleton-Century-Crofts.

Berry, L. L., Carbone, L. P., & Haeckel, S. H. (2002). Managing the Total Customer Experience. *MIT Sloan Management Review, 43*(3), 85-89.

Berry, L. L., & Lampo, S. K. (2000). Teaching an Old Service New Tricks: The Promise of Service Redesign. *Journal of Service Research, 2*(3), 265-275.

Berry, L. L., Wall, E. A., & Carbone, L. P. (2006). Service Clues and Customer Assessment of the Service Experience: Lessons from Marketing. *Academy of Management Perspectives, 20*(2), 43-57.

Binder, M. D., Hirokawa, N., & Windhorst, U. (2009). *Encyclopedia of Neuroscience*. Berlin u.a.: Springer.

Bitner, M. J. (1990). Evaluating Service Encounters: The Effects of Physical Surroundings and Employee Responses. *Journal of Marketing, 54*(2), 69-82.

Bitner, M. J. (1992). Servicescapes: The Impact of Physical Surroundings on Customers and Employees. *Journal of Marketing, 56*(2), 57-71.

Bitner, M. J. (2000). The Servicescape. In T. A. Swartz (Ed.), *Handbook of Services Marketing & Management* (1. ed., pp. 37-50). Thousand Oaks, CA: SAGE Publications.

Bitner, M. J., Booms, B. H., & Mohr, L. A. (1994). Critical Service Encounters: The Employee's Viewpoint. *Journal of Marketing, 58*(4), 95-106.

Bolger, N., Davis, A., & Rafaeli, E. (2003). DIARY METHODS: Capturing Life as it is Lived. *Annual Review of Psychology, 54*(1), 579-616.

Bone, P. F., & Jantrania, S. (1992). Olfaction as a Cue for Product Quality. *Marketing Letters, 3*(3), 289-296.

Bosmans, A. (2006). Scents and Sensibility: When Do (In)Congruent Ambient Scents Influence Product Evaluations? *Journal of Marketing, 70*(3), 32-43.

Bradford, K., & Desrochers, D. (2009). The Use of Scents to Influence Consumers: The Sense of Using Scents to Make Cents. *Journal of Business Ethics, 90*(2), 141-153.

Brady, M. K., & Cronin Jr., J. J. (2001). Some New Thoughts on Conceptualizing Perceived Service Quality: A Hierarchical Approach. *Journal of Marketing, 65*(3), 34-49.

Brakus, J. J., Schmitt, B. H., & Zarantonello, L. (2009). Brand Experience: What Is It? How Is It Measured? Does It Affect Loyalty? *Journal of Marketing, 73*(3), 52-68.

Brand, G., & Millot, J.-L. (2001). Sex differences in human olfaction: Between evidence and enigma. *The Quarterly Journal of Experimental Psychology Section B, 54*(3), 259-270.

Brehm, J. W. (1966). *A theory of psychological reactance.* New York, NY: Academy Press.

Brexendorf, T. O., Mühlmeier, S., Tomczak, T., & Eisend, M. (2010). The impact of sales encounters on brand loyalty. *Journal of Business Research, 63*(11), 1148-1155.

Brooks, C. M. (1987). Leisure Time Physical Activity Assessment of American Adults through an Analysis of Time Diaries Collected in 1981. *American Journal of Public Health, 77*(4), 455-460.

Brüggen, E. C., Foubert, B., & Gremler, D. D. (2011). Extreme Makeover: Short- and Long-Term Effects of a Remodeled Servicescape. *Journal of Marketing, 75*(5), 71-87.

Buck, L. B., & Axel, R. (1991). A Novel Multigene Family May Encode Odorant Receptors: A Molecular Basis for Odor Recognition. *Cell, 65*(1), 175-187.

Bundesamt für Bauwesen und Raumordnung. (2007). Frauen – Männer – Räume: Geschlechterunterschiede in den regionalen Lebensverhältnissen (Vol. 26). Berlin.

Bundesministerium für Umwelt. (2005). *Verbesserung der Luftqualität in Innenräumen - Ausgewählte Handlungsschwerpunkte aus Sicht BMU -.* Retrieved from http://www.apug.de/archiv/pdf/BMU_bericht_innenraumluft_2005.pdf.

Burnes, B., & Cooke, B. (2013). Kurt Lewin's Field Theory: A Review and Re-evaluation. *International Journal of Management Reviews, 15*(4), 408-425.

Burroughs, H. E., & Hansen, S. J. (2004). *Managing indoor air quality* (3. ed.). Lilburn, GA: Fairmont Press.

Campbell, D. T., & Stanley, J. C. (1966). *Experimental and quasi-experimental designs for research.* Boston, MA: Houghton Mifflin.

Capelli, L., Sironi, S., del Rosso, R., & Guillot, J.-M. (2013). Measuring odours in the environment vs. dispersion modelling: A review. *Atmospheric Environment, 79,* 731-743.

Chebat, J.-C., Morrin, M., & Chebat, D.-R. (2009). Does Age Attenuate the Impact of Pleasant Ambient Scent on Consumer Response? *Environment and Behavior, 41*(2), 258-267.

Chen, D., & Dalton, P. (2005). The Effect of Emotion and Personality on Olfactory Perception. *Chemical Senses, 30*(4), 345-351.

Clarke, I., & Schmidt, R. A. (1995). Beyond the servicescape: The experience of place. *Journal of Retailing and Consumer Services, 2*(3), 149-162.

Cohen, J. (1988). *Statistical Power Analysis for the Behavioral Sciences* (2. ed.). Hillsdale, NJ: Lawrence Erlbaum Associates.

Croy, I., Hoffmann, H., Philpott, C., Rombaux, P., Welge-Luessen, A., Vodicka, J., Konstantinidis, I., Morera, E., & Hummel, T. (2014). Retronasal testing of olfactory function: an investigation and comparison in seven countries. *European Archives of Oto-Rhino-Laryngology, 271*(5), 1087-1095.

Dabholkar, P. A., Sheperd, C. D., & Thorpe, D. I. (2000). A Comprehensive Framework for Service Quality: An Investigation of Critical Conceptual and Measurement Issues Through a Longitudinal Study. *Journal of Retailing, 76*(2), 139-174.

Davies, B. J., Kooijman, D., & Ward, P. (2003). The Sweet Smell of Success: Olfaction in Retailing. *Journal of Marketing Management, 19*(5-6), 611-627.

Davis, T. R. V. (1984). The Influence of the Physical Environment in Offices. *Academy of Management Review, 9*(2), 271-283.

DB Vertrieb GmbH. (2014). Willkommen an Bord der Kneipp®-Lechfeld-Bahn! Retrieved 18.11., 2014, from http://www.bahn.de/regio_allgaeu_schwaben/view/wir/teilnetze/kneipp_lechfeld_bahn.shtml

de Lima, A. M., Sapienza, G. B., de Oliveira Giraud, V., & Fragoso, Y. D. (2011). Odors as triggering and worsening factors for migraine in men. *Arquivos de Neuro-Psiquiatria, 69*(2-B), 324-327.

Degel, J., & Köster, E. P. (1999). Odors: Implicit Memory and Performance Effects. *Chemical Senses, 24*(3), 317-325.

Delaunay-El Allam, M., Soussignan, R., Patris, B., Marlier, L., & Schaal, B. (2010). Long-lasting memory for an odor acquired at the mother's breast. *Developmental Science, 13*(6), 849-863.

Deutsche Forschungsgemeinschaft. (2012). *Leitfaden für die Antragstellung Projektanträge.* Bonn: Retrieved from http://www.dfg.de/formulare/54_01/.

Deutsche Forschungsgemeinschaft. (2013). *Sicherung guter wissenschaftlicher Praxis / Safeguarding Good Scientific Practice.* Weinheim: WILEY-VCH. Retrieved from http://www.dfg.de/download/pdf/dfg_im_profil/reden_stellungnahmen/download/empfehlung_wiss_praxis_1310.pdf.

Dick, A. S., & Basu, K. (1994). Customer Loyalty: Toward an Integrated Conceptual Framework. *Journal of the Academy of Marketing Science, 22*(2), 99-113.

Diederich, H. (1966). Zur Theorie des Verkehrsbetriebes. *Zeitschrift für Betriebswirtschaft, 36*(Zweites Ergänzungsheft), 37-52.

Doty, R. L. (1991a). Influences of Aging on Human Olfactory Function. In D. G. Laing, R. L. Doty & W. Breipohl (Eds.), *The Human Sense of Smell* (pp. 181-195). Berlin u.a.: Springer.

Doty, R. L. (1991b). Olfactory Function in Neonates. In D. G. Laing, R. L. Doty & W. Breipohl (Eds.), *The Human Sense of Smell* (pp. 155-165). Berlin u.a.: Springer.

Doty, R. L. (1991c). Psychophysical Measurement of Odor Perception in Humans. In D. G. Laing, R. L. Doty & W. Breipohl (Eds.), *The Human Sense of Smell* (pp. 95-134). Berlin u.a.: Springer.

Doty, R. L. (2001). Olfaction. *Annual Review of Psychology, 52*(1), 423-452.

Doty, R. L., Applebaum, S., Zusho, H., & Settle, R. G. (1985). Sex Differences in Odor Identification Ability: A Cross-cultural Analysis. *Neuropsychologia, 23*(5), 667-672.

Doucé, L., Janssens, W., Swinnen, G., & van Cleempoel, K. (2014). Influencing consumer reactions towards a tidy versus a messy store using pleasant ambient scents. *Journal of Environmental Psychology, 40*, 351-358.

Doucé, L., Poels, K., Janssens, W., & de Backer, C. (2013). Smelling the books: The effect of chocolate scent on purchase-related behavior in a bookstore. *Journal of Environmental Psychology, 36*, 65-69.

Douek, E. (1974). *The Sense of Smell and its Abnormalities*. Edinburgh; London: Churchill Livingstone.

Edvardsson, B., Enquist, B., & Johnston, R. (2010). Design Dimensions of Experience Rooms for Service Test Drives: Case Studies in Several Service Contexts. *Managing Service Quality, 20*(4), 312-327.

Elejalde-Ruiz, A. (2014, 18.04.). For branding, many places adopt signature scents. *Los Angeles Times*. Retrieved from http://www.latimes.com/business/la-fi-scent-branding-20140419-story.html#axzz2zXZ6NTqE&page=1.

Elie-Dit-Cosaque, C., Pallud, J., & Kalika, M. (2012). The Influence of Individual, Contextual, and Social Factors on Perceived Behavioral Control of Information Technology: A Field Theory Approach. *Journal of Management Information Systems, 28*(3), 201-234.

Emsenhuber, B. (2011). Scent Marketing: Making Olfactory Advertising Pervasive. In J. Müller, F. Alt & D. Michelis (Eds.), *Pervasive Advertising* (pp. 343-360). London: Springer.

Engen, T. (1972). The Effect of Expectation on Judgments of Odor. *Acta Psychologica, 36*(6), 450-458.

Engen, T. (1991). *Odor Sensation and Memory*. New York, NY: Preager.

Engen, T., Kilduff, R. A., & Rummo, N. J. (1975). The Influence of Alcohol on Odor Detection. *Chemical Senses, 1*(3), 323-329.

Epple, G., & Herz, R. S. (1999). Ambient Odors Associated to Failure Influence Cognitive Performance in Children. *Developmental Psychobiology, 35*(2), 103-107.

Faul, F., Erdfelder, E., Lang, A.-G., & Buchner, A. (2007). G*Power 3: A flexible statistical power analysis program for the social, behavioral, and biomedical sciences. *Behavior Research Methods, 39*(2), 175-191.

Festinger, L. (1957). *A Theory of Cognitive Dissonance.* Stanford, CA: Stanford University Press.

Fichtel, S. (2009). *"What is Beautiful is Good": Impact of Employee Attractiveness on Market Success* (Vol. 68). Munich: FGM-Verlag.

Field, A. (2013). *Discovering Statistics Using IBM SPSS Statistics* (4. ed.). London u.a.: SAGE Publications.

Fiore, A. M., & Kim, S. (1997). Olfactory Cues of Appearance Affecting Impressions of Professional Image of Women. *Journal of Career Development, 23*(4), 247-263.

Fiore, A. M., Yah, X., & Yoh, E. (2000). Effects of a Product Display and Environmental Fragrancing on Approach Responses and Pleasurable Experiences. *Psychology & Marketing, 17*(1), 27-54.

Fisher, J. D. (1974). Situation-Specific Variables as Determinants of Perceived Environmental Aesthetic Quality and Perceived Crowdedness. *Journal of Research in Personality, 8*(2), 177-188.

Fitzgerald Bone, P., & Scholder Ellen, P. (1999). Scents in The Marketplace: Explaining a Fraction of Olfaction. *Journal of Retailing, 75*(2), 243-262.

Folkman, S., Lazarus, R. S., Dunkel-Schetter, C., DeLongis, A., & Gruen, R. J. (1986). Dynamics of a Stressful Encounter: Cognitive Appraisal, Coping, and Encounter Outcomes. *Journal of Personality and Social Psychology, 50*(5), 992-1003.

Fritz, C. O., Morris, P. E., & Richler, J. J. (2012). Effect Size Estimates: Current Use, Calculations, and Interpretation. *Journal of Experimental Psychology: General, 141*(1), 2-18.

Frontczak, M., Schiavon, S., Goins, J., Arens, E., Zhang, H., & Wargocki, P. (2012). Quantitative relationships between occupant satisfaction and satisfaction aspects of indoor environmental quality and building design. *Indoor Air, 22*(2), 119-131.

Frontczak, M., & Wargocki, P. (2011). Literature survey on how different factors influence human comfort in indoor environments. *Building and Environment, 46*(4), 922-937.

Frye, R. E., Schwartz, B. S., & Doty, R. L. (1990). Dose-Related Effects of Cigarette Smoking on Olfactory Function. *The Journal Of The American Medical Association, 263*(9), 1233-1236.

Gaygen, D. E., & Hedge, A. (2008). Effects of Acute Exposure to a Complex Fragrance on Lexical Access. *Chemical Senses, 34*(1), 85-91.

Gilbert, A. N., Knasko, S. C., & Sabini, J. (1997). Sex Differences in Task Performance Associated with Attention to Ambient Odor. *Archives of Environmental Health, 52*(3), 195-199.

Girard, A. (2015). Langfristiger Einsatz von (Raum-)Düften bei Dienstleistern: Eine kritische Diskussion. In A. Meyer (Ed.), *Aktuelle Aspekte in der Dienstleistungsforschung* (pp. 35-64). Wiesbaden: Springer Gabler.

Girard, M. (2015). *Duft als Erfolgsfaktor: Die Wirkung gezielt eingesetzter Duftstoffe im Dienstleistungsumfeld.* Munich: FGM-Verlag.

Girard, M., Girard, A., Meyer, A., Rosenbusch, B., & Müller-Grünow, R. (2013). Markenduft als Treiber der Service Experience. *Marketing Review St. Gallen, 30*(6), 70-80.

Girard, M., Girard, A., & Suppin, A.-C. (2015). *The Scentscape – An Integrative Framework Describing Ambient Scents in the Servicescape.* Working Paper. Institut für Marketing, Ludwig-Maxmilians-University. Munich.

Girden, E. R. (1992). *ANOVA Repeated Measures.* Newbury Park, CA: SAGE Publications.

Global Industry Analysts Inc. (2012). Air Fresheners: A Global Strategic Business Report. Retrieved from http://www.prweb.com/releases/2011/1/prweb8041794.htm.

Goel, N., Hyungsoo, K., & Lao, R. P. (2005). An Olfactory Stimulus Modifies Nighttime Sleep in Young Men and Women. *Chronobiology International, 22*(5), 889-904.

Gold, M. (1992). Metatheory and Field Theory in Social Psychology: Relevance or Elegance? *Journal of Social Issues, 48*(2), 67-78.

Goldkuhl, L., & Styvén, M. (2007). Sensing the scent of service success. *European Journal of Marketing, 41*(11), 1297-1305.

Goris, E., & Hutter, C.-P. (2011). *Der Duft-Code: Wie die Industrie unsere Sinne manipuliert.* Munich: Heyne.

Gottfried, J. A., & Dolan, R. J. (2003). The Nose Smells What the Eye Sees: Crossmodal Visual Facilitation of Human Olfactory Perception. *Neuron, 39*(2), 375-386.

Gould, A., & Martin, G. (2001). 'A Good Odour to Breathe?' The Effect of Pleasant Ambient Odour on Human Visual Vigilance. *Applied Cognitive Psychology, 15*(2), 225-232.

Grabenhorst, F., Rolls, E. T., & Margot, C. (2011). A hedonically complex odor mixture produces an attentional capture effect in the brain. *NeuroImage, 55*(2), 832-843.

Greenwald, A. G. (1976). Within-Subjects Designs: To Use or Not to Use? *Psychological Bulletin, 83*(2), 314-320.

Grewal, D., & Levy, M. (2009). Emerging Issues in Retailing Research. *Journal of Retailing, 85*(4), 522-526.

Gross Sobol, M. (1959). Panel Mortality and Panel Bias. *Journal of the American Statistical Association, 54*(285), 52-68.

Guéguen, N., & Petr, C. (2006). Odors and consumer behavior in a restaurant. *International Journal of Hospitality Management, 25*(2), 335-339.

Guest, G., Bunce, A., & Johnson, L. (2006). How Many Interviews Are Enough?: An Experiment with Data Saturation and Variability. *Field Methods, 18*(1), 59-82.

Gulas, C. S., & Bloch, P. H. (1995). Right Under our Noses: Ambient Scent and Consumer Responses. *Journal of Business and Psychology, 10*(1), 87-98.

Habel, U., Koch, K., Pauly, K., Kellermann, T., Reske, M., Backes, V., Seiferth, N. Y., Stöcker, T., Kircher, T., Amunts, K., Jon Shah, N., & Schneider, F. (2007). The influence of olfactory-induced negative emotion on verbal working memory: Individual differences in neurobehavioral findings. *Brain Research, 1152*, 158-170.

Harris, L. C., & Goode, M. M. H. (2004). The four levels of loyalty and the pivotal role of trust: a study of online service dynamics. *Journal of Retailing, 80*(2), 139-158.

Hatt, H. (2007). Stellungnahme zum Hintergrundpapier des Umweltbundesamtes vom April 2006: „Duftstoffe: Wenn Angenehmes zur Last werden kann" (Im Auftrag von FORUM ESSENZIA e.V. Bochum 08.08.07).

Hawkes, G., Houghton, J., & Rowe, G. (2009). Risk and worry in everyday life: Comparing diaries and interviews as tools in risk perception research. *Health, Risk & Society, 11*(3), 209-230.

Hehn, P. (2007). *Emotionale Markenführung mit Duft, Duftwirkungen auf die Wahrnehmung und Beurteilung von Marken*. Göttingen-Rosdorf: ForschungsForum.

Heinzerling, D., Schiavon, S., Webster, T., & Arens, E. (2013). Indoor environmental quality assessment models: A literature review and a proposed weighting and classification scheme *Building and Environment, 70*, 210-222.

Henion, K. E. (1971). Odor Pleasantness and Intensity: A Single Dimension? *Journal Of Experimental Psychology, 90*(2), 275-279.

Henshaw, V. (2014). *Urban Smellscapes: Understanding and designing city smell environments*. New York, NY: Routledge.

Herrmann, A., Zidansek, M., Sprott, D. E., & Spangenberg, E. R. (2013). The Power of Simplicity: Processing Fluency and the Effects of Olfactory Cues on Retail Sales. *Journal of Retailing, 89*(1), 30-43.

Herz, R. S. (2009). Aromatherapy Facts and Fictions: A Scientific Analysis of Olfactory Effects on Mood, Physiology and Behavior. *International Journal of Neuroscience, 119*(2), 263-290.

Herz, R. S., & Cupchik, G. C. (1992). An experimental characterization of odor-evoked memories in humans. *Chemical Senses, 17*(5), 519-528.

Herz, R. S., Eliassen, J., Beland, S., & Souza, T. (2004). Neuroimaging evidence for the emotional potency of odor-evoked memory. *Neuropsychologia, 42*(3), 371-378.

Heuberger, E., Hongratanaworakit, T., Böhm, C., Weber, R., & Buchbauer, G. (2001). Effects of Chiral Fragrances on Human Autonomic Nervous System Parameters and Self-evaluation. *Chemical Senses, 26*(3), 281-292.

Heuberger, E., Redhammer, S., & Buchbauer, G. (2004). Transdermal Absorption of (–)-Linalool Induces Autonomic Deactivation but has No Impact on Ratings of Well-Being in Humans. *Neuropsychopharmacology, 29*(10), 1925-1932.

Hirsch, A. R. (1995). Effects of Ambient Odors on Slot-Machine Usage in a Las Vegas Casino. *Psychology & Marketing, 12*(7), 585-594.

Hirschman, A. O. (1970). *Exit, Voice, and Loyalty: Responses to Decline in Firms, Organizations, and States*. Cambridge, MA u.a.: Harvard University Press.

Ho, C., & Spence, C. (2005). Olfactory facilitation of dual-Task performance. *Neuroscience Letters, 389*(1), 35-40.

Höferl, M., Krist, S., & Buchbauer, G. (2006). Chirality Influences the Effects of Linalool on Physiological Parameters of Stress. *Planta Med, 72*(13), 1188-1192.

Homburg, C., & Giering, A. (1996). Konzeptualisierung und Operationalisierung Komplexer Konstrukte: Ein Leitfaden für die Marketingforschung. *Marketing: Zeitschrift für Forschung und Praxis, 18*(1), 5-24.

Homburg, C., Koschate, N., & Hoyer, W. D. (2006). The Role of Cognition and Affect in the Formation of Customer Satisfaction: A Dynamic Perspective. *Journal of Marketing, 70*(3), 21-31.

Houston, M. B., Bettencourt, L. A., & Wenger, S. (1998). The Relationship Between Waiting in a Service Queue and Evaluations of Service Quality: A Field Theory Perspective. *Psychology & Marketing, 15*(8), 735-753.

Huynh, H., & Feldt, L. S. (1976). Estimation of the Box Correction for Degrees of Freedom from Sample Data in Randomized Block and Split-Plot Designs. *Journal of Educational Statistics, 1*(1), 69-82.

International Fragrance Association. (2006). IFRA Code of Practice. Retrieved from http://www.ifraorg.org/en-us/code_of_practice_1.

Jacob, C., Stefan, J., & Guéguen, N. (2014). Ambient scent and consumer behavior: a field study in a florist's retail shop. *The International Review of Retail, Distribution and Consumer Research, 24*(1), 116-120.

Javalgi, R. G., & Moberg, C. R. (1997). Service loyalty: implications for service providers. *Journal of Services Marketing, 11*(3), 165-179.

Jellinek, P. (1996). *Psychologiscal Basis of Perfumery* (4. ed.). London u.a.: Chapman & Hall.

Jones, A. P. (1999). Indoor air quality and health. *Atmospheric Environment, 33*(28), 4535-4564.

Kahneman, D., Krueger, A. B., Schkade, D., Schwarz, N., & Stone, A. (2004a). The Day Reconstruction Method (DRM): Instrument Documentation. Retrieved 15.07., 2014, from http://sitemaker.umich.edu/norbert.schwarz/files/drm_documentation_july_2004.pdf.

Kahneman, D., Krueger, A. B., Schkade, D., Schwarz, N., & Stone, A. (2004b). A Survey Method for Characterizing Daily Life Experience: The Day Reconstruction Method. *Science, 306*(3), 1776-1780.

Kahneman, D., & Tversky, A. (1979). Prospect Theory: An Analysis of Decision under Risk. *Econometrica, 47*(2), 263-291.

Kanji, G. K. (2006). *100 Statistical Tests* (3. ed.). Los Angeles, CA u.a.: SAGE Publications.

Kassarjian, H. H. (1973). Field Theory in Consumer Behavior. In S. Ward & T. S. Robertson (Eds.), *CONSUMER BEHAVIOR: Theoretical Sources* (pp. 118-140). Englewood Cliffs, NJ: Prentice Hall.

Kassarjian, H. H. (1977). Content Analysis in Consumer Research. *Journal of Consumer Research, 4*(1), 8-18.

Keaveney, S. M. (1995). Customer Switching Behavior in Service Industries: A Exploratory Study. *Journal of Marketing, 59*(2), 71-82.

Keller, A., Hempstead, M., Gomez, I. A., Gilbert, A. N., & Vosshall, L. B. (2012). An olfactory demography of a diverse metropolitan population. *BMC Neuroscience, 13*(1), 122-138.

Kelley, H. H., & Michela, J. L. (1980). Attribution Theory and Research. *Annual Review of Psychology, 31*, 457-501.

Keselman, H. J., & Algina, J. (1996). The Analysis of Higher-order Repeated Measures Designs. In B. Thompson (Ed.), *Advances in Social Science Methodology* (Vol. 4, pp. 45-70). Greenwich, CT: JAI Press.

Kim, J., & de Dear, R. (2012). Nonlinear relationships between individual IEQ factors and overall workspace satisfaction. *Building and Environment, 49*, 33-40.

Kirk-Smith, M. D., & Booth, D. A. (1987). Chemoreception in human behaviour: experimental analysis of the social effects of fragrances. *Chemical Senses, 12*(1), 159-166.

Klara, R. (2012, 05.03.). Something in the Air: In a growing trend, retailers are perfuming stores with near-subliminal scents. Call it branding's final frontier. *Adweek.* Retrieved from http://www.adweek.com/news/advertising-branding/something-air-138683.

Knasko, S. C. (1992). Ambient odor's effect on creativity, mood, and perceived health. *Chemical Senses, 17*(1), 27-35.

Knasko, S. C. (1993). Performance, Mood, and Health During Exposure in Intermittent Odors. *Archives of Environmental Health, 48*(5), 305-308.

Knasko, S. C. (1995). Pleasant Odors and Congruency: Effects on Approach Behavior. *Chemical Senses, 20*(5), 479 - 487.

Knasko, S. C., Gilbert, A. N., & Sabini, J. (1990). Emotional State, Physical Well-Being, and Performance in the Presence of Feigned Ambient Odor. *Journal of Applied Social Psychology, 20*(16), 1345-1357.

Knoblich, H., Scharf, A., & Schubert, B. (2003). *Marketing mit Duft* (4. ed.). Munich; Vienna: Oldenburg Wissenschaftsverlag.

Kolbe, R. H., & Burnett, M. S. (1991). Content-Analysis Research: An Examination of Applications with Directives for Improving Research Reliability and Objectivity. *Journal of Consumer Research, 18*(2), 243-250.

Köster, E. P., & de Wijk, R. A. (1991). Olfactory Adaptation. In D. G. Laing, R. L. Doty & W. Breipohl (Eds.), *The Human Sense of Smell* (pp. 199-216). Berlin u.a.: Springer.

Krippendorff, K. (1980). *Content Analysis: An Introduction to Its Methodology.* Newbury Park, CA: SAGE Publications.

Krishna, A. (2012). An integrative review of sensory marketing: Engaging the senses to affect perception, judgment and behavior. *Journal of Consumer Psychology, 22*(3), 332-351.

Krishna, A., Elder, R. S., & Caldara, C. (2010). Feminine to smell but masculine to touch? Multisensory congruence and its effect on the aesthetic experience. *Journal of Consumer Psychology, 20*(4), 410-418.

Kroeber-Riel, W., & Weinberg, P. (2003). *Konsumentenverhalten* (8. ed.). Muninch: Vahlen.

Kunz, W. H., & Hogreve, J. (2011). Toward a deeper understanding of service marketing: The past, the present, and the future. *International Journal of Research in Marketing, 28*(3), 231-247.

Kuzel, A. J. (1999). Sampling in Qualitative Inquiry. In B. F. Crabtree & W. L. Miller (Eds.), *Doing Qualitative Research* (2. ed., pp. 33-45). Thousand Oaks, CA: SAGE Publications.

La Buissonnière-Ariza, V., Lepore, F., Kojok, K. M., & Frasnelli, J. (2013). Increased Odor Detection Speed in Highly Anxious Healthy Adults. *Chemical Senses, 38*(7), 577-584.

Laird, D. A. (1935). What Can You Do With Your Nose? *Scientific Monthly, New York, 41*(2), 126-130.

Lam, S. Y. (2001). The Effects of Store Environment on Shopping Behaviors: A Critical Review. *Advances in Consumer Research, 28*(1), 190-197.

Larsson, M., Finkel, D., & Pedersen, N. L. (2000). Odor Identification. *Journal of Gerontology: PSYCHOLOGICAL SCIENCES, 55B*(5), P304-P310.

LaTour, S. A., & Miniard, P. W. (1983). The Misuse of Repeated Measures Analysis in Marketing Research. *Journal of Marketing Research, 20*(1), 45-57.

Lazarus, R. S. (1993). FROM PSYCHOLOGICAL STRESS TO EMOTIONS: A History of Changing Outlooks. *Annual Review of Psychology, 44*, 1-21.

Lazarus, R. S., & Folkman, S. (1984). *Stress, Appraisal, and Coping.* New York, NY: Springer

Lee, S. W. S., & Schwarz, N. (2012). Bidirectionality, Mediation, and Moderation of Metaphorical Effects: The Embodiment of Social Suspicion and Fishy Smells. *Journal of Personality and Social Psychology, 103*(5), 737-749.

Legrum, W. (2011). *Riechstoffe, zwischen Gestank und Duft: Vorkommen, Eigenschaften und Anwendung von Riechstoffen und deren Gemischen* (1. ed.). Wiesbaden: Vieweg+Teubner Verlag, Springer.

Lehrner, J. P., Eckersberger, C., Walla, P., Pötsch, G., & Deecke, L. (2000). Ambient odor of orange in a dental office reduces anxiety and improves mood in female patients. *Physiology & Behavior, 71*(1-2), 83-86.

Lehrner, J. P., Marwinski, G., Lehr, S., Johren, P., & Deecke, L. (2005). Ambient odors of orange and lavender reduce anxiety and improve mood in a dental office. *Physiology & Behavior, 86*(1-2), 92-95.

Levine, T. R., & Hullett, C. R. (2002). Eta Squared, Partial Eta Squared, and Misreporting of Effect Size in Communication Research. *Human Communication Research, 28*(4), 612-625.

Levy, L. M., Henkin, R. I., Hutter, A., Lin, C. S., Martins, D., & Schellinger, D. (1997). Functional MRI of Human Olfaction. *Journal of Computer Assisted Tomography, 21*(6), 849-856.

Lewin, K. (1942). Field Theory and Learning. In D. Cartwright (Ed.), *Field Theory in Social Science: Selected Theoretical Papers by Kurt Lewin (1951)* (pp. 60-86). New York, NY: Harper & Brothers.

Lewin, K. (1943). Defining the 'Field at a Given Time'. *Psychological Review, 50*(3), 292-310.

Lewin, K. (1946). Behavior and Development as a Function of the Total Situation. In D. Cartwright (Ed.), *Field Theory in Social Science: Selected Theoretical Papers by Kurt Lewin (1951)* (pp. 238-303). New York, NY: Harper & Brothers

Lewin, M. A. (1998). Kurt Lewin: His Psychology and a Daughter's Recollections. In G. A. Kimble & M. Wertheimer (Eds.), *Portraits of Pioneers in Psychology* (Vol. III, pp. 104-118). Washington, DC; Mahwah, NJ: American Psychological Association and Lawrence Erlbaum Associates.

Li, W., Moallem, I., Paller, K. A., & Gottfried, J. A. (2007). Subliminal Smells Can Guide Social Preferences. *Psychological Science, 18*(12), 1044-1049.

Lindstrom, M. (2005a). *Brand Sense: Build Powerful Brands Through Touch, Taste, Smell, Sight, and Sound.* New York: Free Press.

Lindstrom, M. (2005b). Broad sensory branding. *Journal of Product & Brand Management, 14*(2), 84-87.

Lorig, T. S., Huffman, E., DeMartino, A., & DeMarco, J. (1991). The effects of low concentration odors on EEG and behavior. *Journal of Psychophysiology, 5*(1), 69-77.

Ludvigson, H. W., & Rottman, T. R. (1989). Effects of ambient odors of lavender and cloves on cognition, memory, affect and mood. *Chemical Senses, 14*(4), 525-536.

Lück, H. E. (1996). *Die Feldtheorie und Kurt Lewin - Eine Einführung.* Weinheim: Psychologie Verlags Union.

Lunardo, R. (2012). Negative effects of ambient scents on consumers' skepticism about retailer's motives. *Journal of Retailing and Consumer Services, 19*(2), 179-185.

Lunardo, R., & Mbengue, A. (2013). When atmospherics lead to inferences of manipulative intent: Its effects on trust and attitude. *Journal of Business Research, 66*(7), 823-830.

Lundström, J. N., Mathe, A., Schaal, B., Frasnelli, J., Nitzsche, K., Gerber, J., & Hummel, T. (2013). Maternal status regulates cortical responses to the body odor of newborns. *Frontiers in Psychology, 4*(597), 1-20.

MacInnis, D. J. (2011). A Framework for Conceptual Contributions in Marketing. *Journal of Marketing, 75*(4), 136-154.

Maille, V. (2006). Ambient Scents in Government Offices: Direct and Indirect Effects and Moderating Variables. *Latin American Advances in Consumer Research, 1*, 109-116.

Mainland, J. D., Keller, A., Li, Y. R., Zhou, T., Trimmer, C., Snyder, L. L., Moberly, A. H., Adipietro, K. A., Liu, W. L. L., Zhuang, H., Zhan, S., Lee, S. S., Lin, A., & Matsunami, H. (2014). The missense of smell: functional variability in the human odorant receptor repertoire. *Nature Neuroscience, 17*(1), 114-120.

Mandler, G. (1982). The Structure of Value – Accounting for Taste. In S. T. Fiske & M. S. Clark (Eds.), *Affect and Cognition* (pp. 3-36). Hillsdale, NJ: Lawrence Erlbaum.

Mani, G. (1999). *Smells and Multimodal Learning: The Role of Congruency in the Processing of Olfactory, Visual and Verbal elements of Product Offerings.* Ann Arbor, MI: UMI Dissertation Services.

Marchand, S., & Arsenault, P. (2002). Odors modulate pain perception: A gender-specific effect. *Physiology & Behavior, 76*(2), 251-256.

Masago, R., Matsuda, T., Kikuchi, Y., Miyazaki, Y., Iwanaga, K., Harada, H., & Katsuura, T. (2000). Effects of Inhalation of Essential Oils on EEG Activity and Sensory Evaluation. *Journal of physiological Anthropology and Applied Human Science, 19*(1), 35-42.

Mason, M. (2010). Sample Size and Saturation in PhD Studies Using Qualitative Interviews. *Forum: Qualitative Social Research, 11*(3), Art. 8.

Mattila, A. S., & Wirtz, J. (2001). Congruency of scent and music as a driver of in-store evaluations and behavior. *Journal of Retailing, 77*(2), 273-289.

Mayring, P. (2000). Qualitative Content Analysis. *Forum: Qualitative Social Research, 1*(2), 105-114.

McClelland, D. C., Atkinson, J. W., Clark, R. A., & Lowell, E. L. (1976). *The Achievement Motive* (reissued ed.). New York, NY: Irvington.

McDonnell, J. (2007). Music, scent and time preferences for waiting lines. *International Journal of Bank Marketing, 25*(4), 223-237.

McGlone, F., Österbauer, R. A., Demattè, L. M., & Spence, C. (2013). The Crossmodal Influence of Odor Hedonics on Facial Attractiveness: Behavioural and fMRI Measures. In F. Signorelli & D. Chirchiglia (Eds.), *Functional Brain Mapping and the Endeavor to Understand the Working Brain* (pp. 209-225). Rijeka: InTech.

McLellan, E., MacQueen, K. M., & Neidig, J. L. (2003). Beyond the Qualitative Interview: Data Preparation and Transcription. *Field Methods, 15*(1), 63-84.

Mehrabian, A. (1995). Theory and Evidence Bearing on a Scale of Trait Arousability. *Current Psychology, 14*(1), 3-28.

Mehrabian, A., & Russell, J. A. (1974). *An Approach to Environmental Psychology*. Cambridge, MA: The MIT Press.

Meyer, A. (1991). Dienstleistungs-Marketing. *Die Betriebswirtschaft, 51*(2), 195-209.

Meyer, A. (1996). *Dienstleistungsmarketing - Erkenntnisse und praktische Beispiele* (7., unveränd. ed. Vol. 20). München: FGM-Verlag.

Meyer, A., & Mattmüller, R. (1987). Qualität von Dienstleistungen: Entwurf eines praxisorientierten Qualitätsmodells. *Marketing: Zeitschrift für Forschung und Praxis, 9*(3), 187-195.

Meyer, A., & Oevermann, D. (1995). Kundenbindung. In B. Tietz, R. Köhler & J. Zentes (Eds.), *Handwörterbuch des Marketing* (pp. 1340-1351). Stuttgart: Schäffer-Poeschel.

Michon, R., Chebat, J.-C., & Turley, L. W. (2005). Mall atmospherics: the interaction effects of the mall environment on shopping behavior. *Journal of Business Research, 58*(5), 576-583.

Miles, A. N., & Berntsen, D. (2011). Odour-induced mental time travel into the past and future: Do odour cues retain a unique link to our distant past? *Memory, 19*(8), 930-940.

Mitchell, D. J., Kahn, B. E., & Knasko, S. C. (1995). There's Something in the Air: Effects of Congruent or Incongruent Ambient Odor on Consumer Decision Making. *Journal of Consumer Research, 22*(2), 229-238.

Møller, P., & Dijksterhuis, G. (2003). Differential human electrodermal responses to odours. *Neuroscience Letters, 346*(3), 129-132.

Morrin, M., & Chebat, J.-C. (2005). Person-Place Congruency: The Interactive Effects of Shopper Style and Atmospherics on Consumer Expenditures *Journal of Service Research, 8*(2), 181-191.

Morrin, M., & Ratneshwar, S. (2000). The Impact of Ambient Scent on Evaluation, Attention, and Memory for Familiar and Unfamiliar Brands. *Journal of Business Research, 49*(2), 157-165.

Morrin, M., & Ratneshwar, S. (2003). Does It Make Sense to Use Scents to Enhance Brand Memory? *Journal of Marketing Research, 40*(1), 10-25.

Morrison, M., Gan, S., Dubelaar, C., & Oppewal, H. (2011). In-store music and aroma influences on shopper behavior and satisfaction. *Journal of Business Research, 64*(6), 558-564.

Mücke, W., & Lemmen, C. (2010). *Duft und Geruch: Wirkung und gesundheitliche Bedeutung von Geruchsstoffen* (1. ed.). Heidelberg; Munich u.a.: ecomed.

Nemec, A. F. L. (1996). *Analysis of Repeated Measures and Time Series: An Introduction with Forestry Examples, Biometrics Information Handbook No. 6.* Victoria, B.C.: Province of British Columbia Ministry of Forests Research Program. Retrieved from http://www.for.gov.bc.ca/hfd/pubs/docs/wp/wp15.pdf.

Neslin, S. A., Grewal, D., Leghorn, R., Shankar, V., Teerling, M. L., Thomas, J. S., & Verhoef, P. C. (2006). Challenges and Opportunities in Multichannel Customer Management. *Journal of Service Research, 9*(2), 95-112.

Nischk, F. (Writer). (2013, 09.07.). Dufte Bahn - Der Duft für zufriedene Fahrgäste [Television], *Quarks & Co.* Cologne: WDR.

Norbäck, D. (2009). An update on sick building syndrome. *Current Opinion in Allergy and Clinical Immunology, 9*(1), 55-59.

Nordin, S., Claeson, A.-S., Andersson, M., Sommar, L., Andrée, J., Lundqvist, K., & Andersson, L. (2013). Impact of Health-Risk Perception on Odor Perception and Cognitive Performance. *Chemosensory Perception, 6*(4), 190-197.

Olahut, M. R. (2013). The Effects Of Ambient Scent On Consumer Behavior: A Review Of The Literature. *Annals of Faculty of Economic, 1*(1), 1797-1806.

Olejnik, S., & Algina, J. (2003). Generalized Eta and Omega Squared Statistics: Measures of Effect Size for Some Common Research Designs. *Psychological Methods, 8*(4), 434-447.

Orth, U. R., & Bourrain, A. (2008). The influence of nostalgic memories on consumer exploratory tendencies: Echoes from scents past. *Journal of Retailing and Consumer Services, 15*(4), 277-287.

Owen, C. M., & Patterson, J. (2002). Odour liking physiological indices: a correlation of sensory and electrophysiological responses to odour. *Food Quality and Preference, 13*(5), 307-316.

Parish, J. T., Berry, L. L., & Lam, S. Y. (2008). The Effect of the Servicescape on Service Workers. *Journal of Service Research, 10*(3), 220-238.

Parsons, A. G. (2009). Use of Scent in a Naturally Odourless Store. *International Journal of Retail & Distribution Management, 37*(5), 440-452.

Pause, B. M., Ferstl, R., & Fehm-Wolfsdorf, G. (1998). Personality and Olfactory Sensitivity. *Journal of Research in Personality, 32*(4), 510-518.

Perreault Jr., W. D., & Leigh, L. E. (1989). Reliability of Nominal Data Based on Qualitative Judgments. *Journal of Marketing Research, 26*(2), 135-148.

Petzold, K. (1983). *Raumlufttemperatur* (2. ed.). Berlin: VEB Verlag Technik.

Pfaff, D. W. (2006). *Brain Arousal and Information Theory.* Cambridge, MA; London: Harvard University Press.

Poellinger, A., Thomas, R., Lio, P., Lee, A., Makris, N., Rosen, B. R., & Kwong, K. K. (2001). Activation and Habituation in Olfaction—An fMRI Study. *NeuroImage, 13*(4), 547-560.

Proust, M. (1913). *Du côté de chez Swann (Swann's Way).* Paris: Grasset.

Pyrski, M., & Zufall, F. (2009). Odor. In M. D. Binder, N. Hirokawa & U. Windhorst (Eds.), *Encyclopedia of Neuroscience.* Berlin u.a.: Springer.

Qu, S. Q., & Dumay, J. (2011). The qualitative research interview. *Qualitative Research in Accounting & Management, 8*(3), 238-264.

Raudenbush, B., Koon, J., Smith, J., & Zoladz, P. (2003). Effects of Odorant Administration on Objective and Subjective Measures of Sleep Quality, Post-Sleep Mood and Alertness, and Cognitive Performance. *North American Journal of Psychology, 5*(2), 181.

Raudenbush, B., Meyer, B., & Eppich, B. (2002). The Effects of Odors on Objective and Subjective Measures of Athletic Performance. *International Sports Journal, 26*, 14-27.

Ravn, K. (2007, 20.08.). Sniff... and spend. *Los Angeles Times.* Retrieved from http://articles.latimes.com/2007/aug/20/health/he-smell20.

Rempel, J. E. (2006). *Olfaktorische Reize in der Markenkommunikation.* Wiesbaden: Deutscher Universitäts-Verlag.

Robin, O., Alaoui-Ismaïli, O., Dittmar, A., & Vernet-Maury, E. (1999). Basic Emotions Evoked by Eugenol Odor Differ According to the Dental Experience. A Neurovegetative Analysis. *Chemical Senses, 24*(3), 327-335.

Rosenbaum, M. S., & Massiah, C. (2011). An expanded servicescape perspective. *Journal of Service Management, 22*(4), 471-490.

Rotton, J. (1983). Affective and Cognitive Consequences of Malodorous Pollution. *Basic and Applied Social Psychology, 4*(2), 171-191.

Russell, J. A., & Lanius, U. F. (1984). Adaptation Level and the Affective Appraisal of Environments. *Journal of Environmental Psychology, 4*(2), 119-135.

Sakamoto, R., Minoura, K., Usui, A., Ishizuka, Y., & Kanba, S. (2005). Effectiveness of Aroma on Work Efficiency: Lavender Aroma during Recesses Prevents Deterioration of Work Performance. *Chemical Senses, 30*(8), 683-691.

Sakamoto, Y., Ebihara, S., Ebihara, T., Tomita, N., Toba, K., Freeman, S., Arai, H., & Kohzuki, M. (2012). Fall Prevention Using Olfactory Stimulation with Lavender Odor in Elderly Nursing Home Residents: A Randomized Controlled Trial. *Journal of the American Geriatrics Society, 60*(6), 1005-1011.

Sandoz, J. C. (2009). Olfactory Plasticity. In M. D. Binder, N. Hirokawa & U. Windhorst (Eds.), *Encyclopedia of Neuroscience.* Berlin u.a.: Springer.

Scent Marketing Institute. (2011). Scent Marketing Institute: What we do. Retrieved 6.10., 2014, from http://www.scentmarketing.org/about/.

SCENTCOMMUNICATION. (2007). Material Safety Data Sheet: Relax (pp. 1-5). Cologne: aerome GmbH Ambient Air Care.

ScentWorld Marketing. (2013). abous Us I ScentWorld Events. Retrieved 6.10., 2014, from http://www.scentworldevents.com/about_us.

Schenker, S. (2001). Gruesome gourmets. *Nutrition Bulletin, 26*(1), 11-12.

Schifferstein, H. J., Talke, K. S., & Oudshoorn, D.-J. (2011). Can Ambient Scent Enhance the Nightlife Experience? *Chemosensory Perception, 4*(1-2), 55-64.

Schneider, C., Ziemssen, T., Schuster, B., Seo, H.-S., Haehner, A., & Hummel, T. (2009). Pupillary responses to intranasal trigeminal and olfactory stimulation. *Journal of Neural Transmission, 116*(7), 885-889.

Schnell, R., Hill, P. B., & Esser, E. (2005). *Methoden der empirischen Sozialforschung* (7. ed.). Munich; Vienna: Oldenbourg Wissenschaftsverlag.

Schön, M., & Hübner, R. (1996). *Geruch: Messung und Beseitigung* (1. ed.). Würzburg: Vogel.

Schwarz, N. (2004). Metacognitive Experiences in Consumer Judgment and Decision Making. *Journal of Consumer Psychology, 14*(4), 332-348.

Schweizer, C., Edwards, R. D., Bayer-Oglesby, L., Gauderman, W. J., Ilacqua, V., Juhani Jantunen, M., Lai, H. K., Nieuwenhuijsen, M., & Künzli, N. (2007). Indoor time–microenvironment–activity patterns in seven regions of Europe. *Journal of Exposure Science and Environmental Epidemiology, 17*(2), 170-181.

Shadish, W. R., Cook, T. D., & Campbell, D. T. (2002). *Experimental and Quasi-experimental Designs for Generalized Causal Inference.* Boston, MA u.a.: Houghton Mifflin.

Shostack, G. L. (1985). Planning the Service Encounter. In J. Czepiel, M. Solomon & C. F. Surprenant (Eds.), *The Service Encounter: Managing Employee/Customer Interaction in Service Businesses* (pp. 243-253). Lexington, MA: Lexington Books.

Simons, G., & Parkinson, B. (2009). Time-dependent observational and diary methodologies and their use in studies of social referencing and interpersonal emotion regulation. *21st Century Society: Journal of the Academy of Social Sciences, 4*(2), 175-186.

Slotnick, B. M., & Weiler, E. (2009). Olfactory Perception. In M. D. Binder, N. Hirokawa & U. Windhorst (Eds.), *Encyclopedia of Neuroscience* (pp. 3007-3010). Berlin u.a.: Springer.

Sorokowska, A., Sorokowski, P., Hummel, T., & Huanca, T. (2013). Olfaction and Environment: Tsimane' of Bolivian Rainforest Have Lower Threshold of Odor Detection Than Industrialized German People. *PLoS One, 8*(7), e69203.

Spangenberg, E. R., Crowley, A. E., & Henderson, P. W. (1996). Improving the Store Environment: Do Olfactory Cues Affect Evaluations and Behaviors? *Journal of Marketing, 60*(2), 67-80.

Spangenberg, E. R., Grohmann, B., & Sprott, D. E. (2005). It's beginning to smell (and sound) a lot like Christmas: the interactive effects of ambient scent and music in a retail setting. *Journal of Business Research, 58*(11), 1583-1589.

Spangenberg, E. R., Sprott, D. E., Grohmann, B., & Tracy, D. L. (2006). Gender-congruent ambient scent influences on approach and avoidance behaviors in a retail store. *Journal of Business Research, 59*(12), 1281-1287.

Specht, N., Fichtel, S., & Meyer, A. (2007). Perception and attribution of employees' effort and abilities: The impact on customer encounter satisfaction. *International Journal of Service Industry Management, 18*(5), 534-554.

Srnka, K. J. (2007). Integration qualitativer und quantitativer Forschungsmethoden: Der Einsatz integrierter Forschungsdesigns als Möglichkeit zur Theorieentwicklung in der Marketingforschung als betriebswirtschaftliche Disziplin. *Marketing: Zeitschrift für Forschung und Praxis, 29*(4), 247-260.

Steenkamp, J.-B. E. M., & Baumgartner, H. (1992). The Role of Optimum Stimulation Level in Exploratory Consumer Behavior. *Journal of Consumer Research, 19*(3), 434-448.

Stroh, K. (2005). *Gerüche und Geruchsbelästigung*. Augsburg: Bayerisches Landesamt für Umwelt. Retrieved from http://www.lfu.bayern.de/ umweltwissen/doc/uw_23_geruchsbelaestigungen.pdf.

Strous, R. D., & Shoenfeld, Y. (2006). To smell the immune system: Olfaction, autoimmunity and brain involvement. *Autoimmunity Reviews, 6*(1), 54-60.

Suh, J.-C., & Yi, Y. (2006). When Brand Attitudes Affect the Customer Satisfaction-Loyalty Relation: The Moderating Role of Product Involvement. *Journal of Consumer Psychology, 16*(2), 145-155.

Sutton, J. (2011, 19.12.). Scent makers sweeten the smell of commerce. *Reuters.com*. Retrieved from http://www.reuters.com/article/2011/12/19/us-usa-scented-idUSTRE7BI1PF20111219.

Symon, G. (1998). Qualitative Research Diaries. In G. Symon & C. Cassell (Eds.), *Qualitative Methods and Analysis in Organizational Research: A Practical Guide* (pp. 94-117). London: SAGE Publications.

Tafalla, M. (2013). A World Without the Olfactory Dimension. *The Anatomical Record, 296*(9), 1287-1296.

Takagi, S. F. (1989). *Human Olfaction*. Tokyo: University of Tokio Press.

Tanguma, J. (1999). Analyzing Repeated Measures Designs Using Univariate and Multivariate Methods: A Primer. In B. Thompson (Ed.), *Advances in Social Science Methodology* (Vol. 5, pp. 233-250). Stamford, CT: JAI Press.

Teller, C., & Dennis, C. (2012). The effect of ambient scent on consumers' perception, emotions and behaviour: A critical review. *Journal of Marketing Management, 28*(1-2), 14-36.

Thomaselli, R. (2006). Trends to Watch in 2007. *Advertising Age, 77*(51), 10.

Twardella, D., Matzen, W., Lahrz, T., Burghardt, R., Spegel, H., Hendrowarsito, L., Frenzel, A. C., & Fromme, H. (2012). Effect of classroom air quality on students concentration: results of a cluster-randomized cross-over experimental study. *Indoor Air, 22*(5), 378-387.

Umweltbundesamt. (2006a). *Duftstoffe: Wenn Angenehmes zur Last werden kann*. Retrieved from http://www.umweltdaten.de/publikationen/fpdf-l/3550.pdf.

Umweltbundesamt. (2006b). *Kurzfassung des Berichts zum Forschungsvorhaben "Untersuchung der Bedeutung luftgetragener Kontaktalergene (Typ-IV-Allergene) bei der Entstehung von Kontaktekzemen"*. Retrieved from http://www.umweltdaten.de/publikationen/kontaktallergene.pdf.

van Harreveld, A. P. (2001). From odorant formation to odour nuisance: new definitions for discussing a complex process. *Water Science and Technology, 44*(9), 9-15.

van Toller, S., Behan, J., Howells, P., Kendal-Reed, M., & Richardson, A. (1993). An analysis of spontaneous human cortical EEG activity to odours. *Chemical Senses, 18*(1), 1-16.

Varadarajan, P. R. (2003). Musings on Relevance and Rigor of Scholarly Research in Marketing. *Journal of the Academy of Marketing Science, 70*(1), 269-376.

Vennemann, M. M., Hummel, T., & Berger, K. (2008). The association between smoking and smell and taste impairment in the general population. *Journal of Neurology, 255*(8), 1121-1126.

Verhoef, P. C., Lemon, K. N., Parasuraman, A., Roggeveen, A., Tsiros, M., & Schlesinger, L. A. (2009). Customer Experience Creation: Determinants, Dynamics and Management Strategies. *Journal of Retailing, 85*(1), 31-41.

Villemure, C., Slotnick, B. M., & Bushnell, M. C. (2003). Effects of odors on pain perception: deciphering the roles of emotion and attention. *Pain, 106*(1-2), 101-108.

Vlahos, J. (2007, 09.09.). Scents and Sensibility. *The New York Times*. Retrieved from http://www.nytimes.com/2007/09/09/realestate/keymagazine/909SCENT-txt.html?pagewanted=all&_r=0.

Voelkle, M. C., & McKnight, P. E. (2012). One size fits all? A Monte-Carlo simulation on the relationship between repeated measures (M)ANOVA and latent curve modeling. *Methodology, 8*(1), 23-38.

von Kempski, D. (2002). The Use of Olfactory Stimulants to Improve Indoor Air Quality. *Journal of the Human-Environmental System, 5*(2), 61-68.

von Kempski, D. (2004). Air and well being - a way to more profitability. *KI Luft- und Kältetechnik, 40*(10), 422-426.

von Rosenstiel, L., & Kirsch, A. (1996). *Psychologie der Werbung*. Rosenheim: Komar.

Waddington, K. (2005). Using diaries to explore the characteristics of work-related gossip: Methodological considerations from exploratory multimethod research. *Journal of Occupational and Organizational Psychology, 78*(2), 221-236.

Wang, B.-L., Takigawa, T., Yamasaki, Y., Sakano, N., Wang, D.-H., & Ogino, K. (2008). Symptom definitions for SBS (sick building syndrome) in residential dwellings. *International Journal of Hygiene and Environmental Health, 211*(1-2), 114-120.

Warhol, A. (1975). *The Philosophy of Andy Warhol: From A to M and Back Again.* New York, NY: Harcourt Brace Jovanovich.

Warm, J. S., Dember, W. N., & Parasuraman, R. (1991). Effects of olfactory stimulation on performance and stress in a visual sustained attention task. *Journal of the Society of Cosmetic Chemists, 42*(3), 199-210.

Warren, J. R., & Halpern-Manners, A. (2012). Panel Conditioning in Longitudinal Social Science Surveys. *Sociological Methods & Research, 41*(4), 491-534.

West, B. T., Welch, K. B., & Galecki, A. T. (2007). *Linear Mixed Models: A Practical Guide Using Statistical Software.* Boca Raton, FL: Taylor & Francis Group.

Wheeler, L. (2008). Kurt Lewin. *Social and Personality Psychology Compass, 2/4*, 1638-1650.

Wheeler, L., & Reis, H. T. (1991). Self-Recording of Everyday Life Events: Origins, Types and Uses. *Journal of Personality, 59*(3), 339-354.

Wicklund, R. A. (1974). *Freedom and Reactance.* Pontomac, MD: L. Erlbaum Associates; distributed by the Halsted Press Division, Wiley.

Winneke, G., Both, R., Frechen, F. B., Hangartner, M., Medrow, W., Paduch, M., & H., P. P. (1995). Charakterisierung von Geruchsbelästigung Teil 2: Verknüpfung von ausgesuchten Geruchsparametern im Hinblick auf Belästigungsrelevanz. *Staub – Reinhaltung der Luft, 55*(3), 113-118.

Zalejska-Jonsson, A., & Wilhelmsson, M. (2013). Impact of perceived indoor environment quality on overall satisfaction in Swedish dwellings. *Building and Environment, 63*, 134-144.

Zeithaml, V. A. (1981). How Consumers Evaluation Processes Differ Between Goods and Services. In J. H. Donnelly & W. R. George (Eds.), *Marketing of Services* (pp. 186-190). Chicago, IL: American Marketing Association.

Zeithaml, V. A. (1988). Consumer Perceptions of Price, Quality, and Value: A Means-End Model and Synthesis of Evidence. *Journal of Marketing, 52*(3), 2-22.

Zemke, D. M., & Shoemaker, S. (2007). Scent across a crowded room: Exploring the effect of ambient scent on social interactions. *Hospitality Management, 26*(4), 927-940.

Zemke, D. M., & Shoemaker, S. (2008). A Sociable Atmosphere: Ambient Scent's Effect on Social Interaction. *Cornell Hospitality Quarterly, 49*(3), 317-329.

Zomerdijk, L. G., & Voss, C. A. (2010). Service Design for Experience-Centric Services. *Journal of Service Research, 13*(1), 67-82.